DRT SCI £5 4/24

MIDDLE WORLD

MIDDLE WORLD

The restless heart of matter and life

Mark Haw

Macmillan

London   New York   Melbourne   Hong Kong

First published 2007 by
Macmillan
Houndmills, Basingstoke, Hampshire RG21 6XS and
175 Fifth Avenue, New York, N. Y. 10010
Companies and representatives throughout the world

ISBN-13: 978–1–4039–8603–0
ISBN-10: 1–4039–8603–7

This book is printed on paper suitable for recycling and made from fully managed and sustained forest sources.

A catalogue record for this book is available from the British Library.

A catalog record for this book is available from the Library of Congress.

10   9   8   7   6   5   4   3   2   1
16   15   14   13   12   11   10   09   08   07

Printed and bound in China

CONTENTS

Chapter 1

TINY THINGS THAT NEVER STAND STILL

One day in June, in the year 1827, in rooms in Soho, London, a 54-year-old Scottish botanist named Robert Brown peered into the eyepiece of his microscope.

He had placed a drop of water under the lens. The water contained pollen grains from a flower named *Clarkia pulchella*, a North American relative of the evening primrose family.

What Brown the botanist saw would finally answer a 2,000-year-old riddle: the mystery of what matter is made of – albeit 80 years later, long after Brown's death.

But peering into his microscope that June day, Brown wasn't actually thinking about matter. He had set himself a different question. He was thinking about life.

He was thinking, specifically, about plant sex.

His droplet of water was full of pollen grains. How, he wanted to know, does pollination work? How do these tiny particles spread the message of life from plant to plant? What are they, and where do they go, and what do they do?

Brown spent the whole of that summer peering into his microscope, performing painstaking observations and writing careful notes on what he saw. By the time August's heat faded into September's mellow fruitfulness, he was forced to accept that he had failed. After three months of experiments he was none the wiser about plant sex. Disappointed, he concluded that, surprising and inexplicable though it was, what he'd seen in his microscope had nothing at all to do with life.

He was right.

And he was also wrong.

This is the story of how some tiny balls of pollen, observed so painstakingly by a Scottish botanist in June, July and August of 1827, first changed the way we understand matter – and are now changing the way we understand life.

Plant sex

Brown was always curious about plants. He grew up scouring the Scottish hills for new species. He became a world-renowned botanical expert, though he had no official qualifications in the science. He was a sprightly, lively featured man, always dressed modestly in black coat and trousers, concealing a racy humour under a habitual shy reserve.

Long before Charles Darwin set out on his own momentous voyage of discovery, Brown too had done his share of rummaging about in the far-flung corners of the globe, collecting all manner of previously unknown botanical marvels. But by 1827 his seafaring days were over, and the focus of his curiosity settled on how plants reproduce.

Scientists knew it had something to do with pollen. But no one knew how pollen actually worked. The pollen grains we've all caught on a sleeve or spotted on a bee's foot are actually hollow sacs, containing thousands of much smaller particles. These smaller particles are the tiny messengers that plants use to communicate with each other – to continue the never-ending process of regeneration. These particles are so small – about a thousandth of a millimetre across, less than a fiftieth of the width of a fine hair – that they can only be seen with a microscope.

Brown and his contemporaries knew that the bursting of the large pollen sacs and release of these minuscule internal specks is a vital stage in plant reproduction: it is the way a plant somehow exchanges characteristics with its fellows to produce the next generation. What they didn't know was where the pollen particles go or how they carry the vital trigger to make new life.

Brown, with his microscope, was one of a new breed of botanists that appeared around the beginning of the 19th century. Over nearly three decades of work he had built up a reputation as one of the best botanists of the age. Much of that reputation was based on the traditional trade of the botanist: identifying, describing, cataloguing and mapping the bewildering variety of species covering the globe. But he was also amongst the pioneers of a new approach in plant science. For Brown, cataloguing and classifying weren't enough – he wanted to experiment, to observe, to see for himself.

Over the years Brown built up a formidable parallel reputation as an expert with the lens. Among the many who sought his advice on microscopes, for instance, was that young ship's naturalist Charles Darwin, preparing, in the 1830s, to set out on the *Beagle*.

So on that June day of 1827 Brown popped open a few pollen sacs from a sample of *Clarkia pulchella*, releasing the tiny internal particles into the droplet of water under the lens of his microscope.

He thought he might find out just a little bit more about pollination. In fact, he was heading straight into the heart of matter itself.

'I observed many of them very evidently in motion...'

Everything began with one simple observation. Its consequences would reverberate about the world of science for the next 200 years.

As he peered into the eyepiece of his microscope, Brown expected to see a calm, motionless suspension of particles – pollen sitting quietly in the droplet of water under the lens, waiting patiently for Brown to watch, study, describe and categorize.

He saw immediately that studying these pollen particles was going to be more complicated than that. The pollen particles weren't sitting calmly at all.

They were moving around all over the place. They weren't just moving – they were *dancing*. They were jumping up and down, zigzagging back and forth, hurling themselves about as if in the grip of some invisible, swirling, microscopic hurricane. And the crazy dance went on and on, not a moment's pause. No matter how long Brown watched, the dance of the pollen particles never stopped. They just would *not* stand still.

The one thing scientists of the early 19th century knew was capable of producing such wilful, limitless, apparently random motions was life. Living things. Brown was reminded of the *animalculae*, the protozoa, the tiny marine creatures that, since the earliest days of the microscope, had been known to move like this.

Could these particles of pollen, even though they were so much tinier than even the tiniest known protozoa, really be *alive*? Were pollen particles the smallest, most fundamental living units, the 'atoms' of life? Was this the secret of plant sex – that pollination was the transmission of the microscopic carriers of life itself?

This was indeed Brown's first idea – that the pollen particles were themselves alive. But he was a careful scientist. It wasn't enough to have an idea and then sit back and marvel at it. Over the following days and weeks Brown carried out further extensive microscopic studies, looking at pollen from a host of other plants – and more importantly, at particles from substances that he knew were not organic. He took grains of mineral dust from rocks and particles of soot. Almost anything he could lay his hands on got ground into small pieces, put into drops of water, and slid under the lens of his microscope.

So then – was the dance of the pollen evidence of tiny atoms of life? Did Brown really have the basic units of life swimming around under the lens of his microscope?

No.

Because, as Brown saw, dead things danced too.

Soot, sand, minerals, metals. Even the deadest dust danced in just the same way as the pollen particles from *Clarkia pulchella*. Everything, as long as it was small enough, danced. The dance of the pollen wasn't evidence of the vital microscopic life force after all: these were not the atoms of life.

Bang went an astounding idea. Not to worry – science is full of temporary disappointments. *Something* was making the pollen and the dust and the soot dance tirelessly around like that, even if it wasn't biology. And whatever it was it was still interesting: Brown had still done three solid months' work and collected a substantial set of careful observations... so, as scientists have done throughout history, negative results or no, Brown published.

His account of the experiments of that summer of 1827 became a paper in the British journal *Philosophical Magazine*, with the title 'A Brief Account of Microscopical Observations made in the Months of June, July, August 1827, on the Particles Contained in the Pollen of Plants; and on the General Existence of Active Molecules in Organic and Inorganic Bodies'.

Despite an initial flurry of excitement – based on the popular and short-lived misconception that Brown *had* found the 'atoms of life' – the dancing pollen story didn't exactly catapult Brown into instant celebrity. For the next half century or so, his experiments were more or less forgotten, ignored, or, even when someone did remember them, profoundly misunderstood.

The dancing muscle

Fast forward 160 years to the 1990s. From pollen particles – to muscle proteins.

Chemical energy is stored in matter all around us. But left to themselves chemicals aren't going to produce motion. To move, we need muscles.

Walking, lifting and turning are produced by muscles connected to bones – skeletal muscles. A skeletal muscle is a bundle of long fibres made from the protein actin. These fibres are connected through tendons to the bones and joints. When we use a muscle, the actin fibres contract, pulling on the tendons, and the tendons pull on the bones, articulating the skeleton.

What makes the actin fibres contract in the first place? Woven amongst the actin fibres are ranks of specialist protein molecules called myosins. To tense a muscle, lines of myosin proteins chemically grab hold of their nearest actin fibre and pull. It's like a microscopic tug-of-war. In turn, the contracted actin fibres pull the tendons, the tendons pull the bones – and you move, or lift, or stand up, or jump.

Of course, using your muscles requires energy. The myosin proteins need to convert chemically stored energy into force to pull on their local actin fibres. Most biological processes are powered by a molecular fuel called adenosine triphosphate or ATP[1]. Chemical breakdown of ATP results in conversion of the stored chemical energy into more useful forms. A myosin protein uses the chemical energy stored by a molecule of ATP to generate mechanical motion and tug on its nearby actin fibre.

How does this microscopic mechanical process really work? The easiest assumption is that it works pretty much like any simple machine – according to regular mechanical rules, like clockwork. In such a clockwork mechanical picture of the muscle, we give a myosin protein an ATP molecule and it uses the chemical energy produced by 'burning' that molecule to take a single tug on its local actin fibre – hauling itself along the fibre and contracting the actin fibre back in the other direction as it goes. Add together a bundle of these myosin–actin combinations and we have a whole muscle: a concert of myosin proteins hauling on actin fibres, in turn pulling on tendons and bones – the tiny tug-of-war that moves you and me around.

The appealing thing about this clockwork mechanics of the muscle is that it's like a simple everyday machine: put in one fuel unit, take one myosin step, pull one actin fibre. Next fuel unit, next step, next pull... rather like the sort of machine we're familiar with, such as the internal combustion engine in a car where a piston is driven in and out of a cylinder, driving the camshaft

---

1　A typical sedentary person might manufacture and consume a cumulative total of as much as 40 kilos of ATP per day. An energetic runner could use this much in about an hour.

around and around to generate a continuous cycle of force. With such an everyday picture, we seem to be able to understand a complex human process like the muscle protein in the same way that a car mechanic would analyze the workings under the bonnet of a BMW. Maybe life is just a question of scaled-down engineering: taking our big machines and simply downsizing into the world of the tiny, the world of the cell.

The problem with this mechanical muscle picture is that, for all its appealing everyday simplicity – it's wrong.

Scientists – like Robert Brown back in 1827 – are rarely completely satisfied until they actually see things with their own eyes. In the case of muscle proteins (and many other cellular processes for that matter), in the early 1990s researchers began to take advantage of the very latest techniques for directly observing actual molecular behaviour, to try to catch a myosin molecule in the act – a snapshot of the muscle machine in motion.

Among these researchers was a team led by Toshio Yanagida at Osaka University in Japan. Yanagida wanted to measure what really happens when you feed myosin some ATP. To try to understand the chemical details of how the muscle protein machine works, he wanted to see a single myosin molecule hauling itself along an actin fibre. Yanagida's group developed ways to watch the motions of single protein molecules as they were energized by ATP. These include tricks like making molecules fluoresce – so that when you point a laser at them in a microscope they beam out like molecular lighthouses, signalling their positions and motions – and techniques to attach tiny beads to the ends of actin fibres to use as 'handles': light pressure from lasers, focused on the beads, holds the actin fibre in place, while energized myosin molecules step along it.

Yanagida's group got to the stage where they could detect and measure the tiny steps that a single myosin protein took along its actin fibre as they held onto the fibre and fed the myosin protein molecules of ATP.

What they saw threw the muscle-as-clockwork-machine picture out of the window.

When you gave a myosin one ATP fuel molecule, it didn't necessarily respond by taking just one step along the local actin fibre. It might take two, three or even five steps. It might even step backwards – pulling in the wrong direction. The myosin molecule seemed to carry out an almost random dance along its actin fibre, like a jittery tightrope walker.

There was more. The same molecule didn't always do the same thing. Sometimes it danced a few steps forward – sometimes a few steps back. Sometimes it didn't move at all. On comparing different myosin molecules under the same conditions, Yanagida's team yet again obtained almost random results: different molecules took different numbers of steps per fuel unit.

That single example, on its own, might not have seemed so significant. Biology is full of odd things – people and plants are such complex entities that things rarely turn out exactly as you might have expected. But, as scientists have been discovering since, a similar story crops up across all kinds of microscopic biological processes. Restlessness isn't confined to muscles. In many cases, a similar near-randomness, noisiness or unpredictability appears to be fundamental to the cellular processes underlying life, such as the carrier molecules that transport chemicals in the cell, the folding of proteins into their precise functional shapes, the working of enzymes, and even how the DNA molecule does its job. Scientists investigating the basic mechanisms of life have been forced to come to terms with a world that is noisy, random and seemingly unpredictable.

How can life work at all when so many of its microscopic machines seem to be so infected by randomness? What kind of engineer builds a machine like that? Surely there must be a better, more efficient, more reliable way?

But, as scientists have begun to realize, life's machines have no alternative. Resistance is useless. Because restlessness is inevitable.

The middle world

What do restless muscles and restless pollen have in common? Size.

Size matters. As Brown showed, it wasn't just pollen particles that danced so unexpectedly and so tirelessly. Grains of minerals and clays did it too. Bits of coal dust (with which 1827 London was liberally supplied), bits of sand – they all danced. The one thing that all Brown's dancing objects had in common was their size.

The smallest were just under a thousandth of a millimetre across. The biggest were perhaps five or ten thousandths of a millimetre. All of them lay within that range, from about a hundredth to about a tenth the size of a human hair. This is a very special size. What happens to objects this size, what happens at this scale of reality, matters to every one of us: it determines what we are and how we function.

It turns out that there are actually a lot of very important objects of this size in our Universe: cells, DNA, viruses, the globules of fat floating in water that go to make milk, the stringy molecules you rub on your head every time you shampoo your hair, and the clever head-and-tail molecules that make soap are just some of them.

The list also includes the restless myosin proteins caught in the act of their jittery tightrope walk by Toshio Yanagida and his team of scientists in Osaka.

This book is the story of the universe of things this size – between a hundredth and a tenth the thickness of a human hair. A universe I've called the *middle world*.

Brown was surprised by the middle world. Before him there had been odd rumours of something strange to be discovered under the microscope, but this unprepossessing Scots botanist was the first to undertake a proper voyage there. As we shall see, most scientists, even the grandest, never imagined there might be important things going on at such a middle scale.

The middle world still suffers from a lack of coverage, almost 200 years after Brown. There are umpteen books about the world of atoms – the weird stuff that goes on at the smallest scales, electrons and nuclei, quarks and superstrings and quantum uncertainty. All this stuff is a lot smaller than a thousandth of a millimetre – a lot smaller than Brown's world of dancing pollen. At the opposite end of the scale, there are shelves devoted to describing the Universe on its grandest scale: galaxies, Big Bangs, black holes and all the other cosmological what-not. A lot bigger than a hundredth of a millimetre – a lot bigger than Brown's world.

One of the messages of this story is that the world in the middle is just as important and interesting a place. There are many scientists working to understand what goes on in the middle world. They often refer to the phenomena they study as *mesoscopic*. *Meso* is a Greek-derived word meaning *middle*: in the middle between the *microscopic* scale of quarks and atoms and the *macroscopic* world of sand, pebbles, people, houses, planets and galaxies.

The middle world isn't just interesting because it doesn't get talked about much. There's something far more important about it: matter behaves differently there. The middle world is special because of one single overwhelming fact about the behaviour of its inhabitants. This is that very first thing that struck Brown the botanist when he first peered down into the lens of his microscope at pollen particles from *Clarkia pulchella*.

Middle world objects *simply cannot stand still*.

The restless heart of matter and life

This book, then, is the story of the middle world, the world where nothing stands still; and the story of its inevitably restless inhabitants: things that, scientists are now learning, lie at the heart of our existence. This book is the story of the restless heart of matter and life.

There's a lot going on in this restless middle world. We are still learning about it, almost 200 years after Brown first visited it in that summer of 1827.

The story of the middle world is a tale that has immense significance to modern-day ideas of matter, technology and life – even though it begins in earnest almost 200 years ago, and has roots that go back to the dawn of Western science in Ancient Greece and perhaps even further – back almost 4 billion years to the origin of life.

It's more than a story of restless pollen and restless proteins. It's a story of all kinds of science and all kinds of people: of accidental discovery, philosophical controversy, mathematical genius and stubborn experimenting. It's a tale of physics, of biology and botany, of chemistry... and of history and people too. A tale of people struggling to understand the world around them: a struggle not just with science but with politics, with philosophies, with prejudices, with ignorance, with disease, with despair and with accident.

It's about ancient Greeks; the last mediæval painter; the first modern scientists; the pioneers of industry; and of course our intrepid explorer–botanist who never went anywhere without his microscope. It's about the third greatest scientist in history; it's about an Austrian whose ideas changed the world, but who finished up hanging from the rafters of a hotel room. It's about Albert Einstein's least known but perhaps most important breakthrough; and the French stock market; and Marie Curie's lover. Nylon stockings, molecular cargo trucks, nanotechnology and washing-up liquid. Oh, and the origin of life.

It's the story behind some of the most exciting science of the present day, science at the borders of our current knowledge about life: how living systems developed the capability to harvest the resources of matter and energy around them.

For what happens in the middle world, as scientists are finally beginning to understand almost two centuries after Brown the botanist, may hold one of the secrets of life. Ironically, Brown had set out to study the great question – How does life work? – by looking at the pollen of *Clarkia pulchella*. And, though he concluded that his observations had nothing to do with life after all, it's beginning to look as if one of the key processes enabling life was exactly what he saw.

The incessant and inevitable restlessness of the middle world.

Chapter 2

THE ACCIDENTAL DISCOVERY OF THE MIDDLE WORLD

On Culloden Moor, a windswept, unsheltered spot east of Inverness just off the road to Nairn and Elgin, there is a collection of gravestones. They are old, lichen-covered, unremarkable. Though each stone features only a single name – some hardly decipherable, worn away by a hundred years of Scottish weather – each one marks the resting place of a large number of men. These are mass graves.

Culloden was the site of the last land battle on mainland British soil. Around noon on a cold April day in 1746, the rebel Highland army of Bonnie Prince Charlie met the Royal army of King George II, led by George's son the Duke of Cumberland. There, in the space of about half an hour, the hopes of the Highland rebellion were crushed.

After the battle the dead were quickly buried. There was little attempt to identify the bodies: the Highlanders were simply separated into families – the famous clans – pushed into large pits, and covered with earth. The small stone memorials weren't even erected until 1881, more than a century later. They mark the clans of the dead.

Under one of the stones lies what little remains of a farmer named John Brown, from the small Scottish town of Forfar. Through the rainy morning of 16 April 1746, this man stood waiting on the moor, in the midst of the second row of Highland columns. Perhaps he had friends, Forfarshire neighbours, about him, perhaps only strangers. He and the men who stood with him belonged to a regiment raised by the 20-year-old Lord Ogilvie, son of the local earl. This regiment of loyal farmers was about to fight for the cause of the Pretender – the true heir to the thrones of Scotland and England.

Perhaps they were proud; perhaps they were frightened. What's certain is that they, along with the rest of the Highlanders, were tired.

The night before, the Highland army had attempted to march up secretly to the encamped Royals in order to deliver a stunning dawn attack. It would have been a daring blow, catching the Royal army by surprise and quite possibly changing the course of history.

If only it hadn't been for the boggy ground. Sloshing around in the mud, losing their way in the dark, the Highlanders did not reach the Royals' camp until well after dawn. By then it was too late. They were too fatigued and dispirited to carry out the attack. The Highlanders were forced to turn around and march right back to Culloden.

On their heels came the Royal Army, commanded by the Duke of Cumberland. By midday the armies were lined up and facing each other across Culloden Moor, barely 300 yards apart. The Royals were fresh and ready, encouraged by the fact that it was they who'd been doing the chasing. The Highlanders were tired, fractious, depressed by the failure of their bold plan.

The sleet blew into the faces of the front ranks of the Scots. They waited for their commander, the Pretender Bonnie Prince Charlie, to issue the order to charge. The stationary Scots suffered mounting casualties as the cannon and grapeshot plunged out of the sky into their ranks. The clans fumed impatiently, desperate to attack – or bickered, tempted to melt away and give up on the rebellion.

Finally Charlie gave the order. This was what the clans did best: hundreds of roaring, charging warriors, flinging swords and daggers, throwing themselves with unflinching courage at the enemy. This terrifying charge had been the Highlanders' secret of success throughout the year-long rebellion. But this time, steeled for a decisive fight, the Royals stood firm. Their artillery continued to decimate the charging Highlanders. Any clansmen that did make it through the first rank of enemy found themselves alone, defenceless. The battle lasted about half an hour. Bodies, the great majority of them Highlanders, littered the windswept field. The rebellion was over.

Records of what happened to the Highlanders are scarce, so we can hardly say for certain just where John Brown from Forfar met his end. As the musket smoke cleared and the cries of the wounded died away, John Brown's body must have been collected up, carried to the corner of the field, and thrown into a mass grave with the rest of Lord Ogilvie's dead.

John Brown did leave something behind. He had a son; that son, James, had a son too. James's son, Robert, would one day be a scientist.

A recipe for a scientist?

Baby Robert was born four days before Christmas 1773. The Browns had set up home in Montrose, a small town on the east coast of Scotland, between Aberdeen to the north and the River Tay to the south. James Brown, son of the dead rebel soldier John, was a stubbornly anti-establishment Minister in the Scottish Episcopal Church, tenacious and ingenious in his efforts to outwit the Penal Laws brought in to quell the last flickerings of Scottish Jacobite resistance.

Undoubtedly something of his father's tenacity and disdain for authority rubbed off on Robert. There was a third ingredient though, beyond independence of mind and ingenuity, essential in a scientist: curiosity. As a young boy, Robert showed plenty of that. He spent his youth roaming the Scottish countryside: not just idly wandering, but carefully examining and observing the natural world around him. He was learning to read the natural encyclopædia of life spread in all its splendour across the Scottish hills.

Robert received a solid education, first at the local grammar school, then at the well-known Marischal College in Aberdeen.

There he studied philosophy and mathematics, and there is evidence that he also followed lectures in botany – the discipline that came to dominate his life. In 1790 the Browns moved to Edinburgh, where Minister James became the leader of a small group of Edinburgh dissenters. Robert followed, taking up studies in medicine at the University of Edinburgh, at that time the world centre of medical science. James Brown's career in Edinburgh didn't last long, however: in 1791 he died of apoplexy.

By this time Robert was already very much a scientist in the making. To be more precise: he was well on the way to becoming a botanist.

A scientific craze

At the end of the 18th century botany was more than a science: it was a popular craze.

Disciplines like molecular biology, genetics, proteomics and biophysics look set to be the sexiest life sciences of the 21st century: the stuff of hi-tech equipment, colourful computer graphics, powerful microscopes and lasers. Botany, by contrast, doesn't seem to get young would-be biologists very excited. But every modern field of biology owes an enormous debt to this more sedate-sounding cousin.

In the late 18th and early 19th centuries, all across Europe, almost everyone who had the wherewithal indulged in a bit of botanizing – professional and amateur alike. For instance, the novelist George Eliot describes in her *Ilfracombe Journal* of 1856 how she and her companions spent their holiday feverishly collecting samples of plants and creatures from rock pools along the Devon coast. Wherever you went there was always a local expert – a parson or a schoolteacher perhaps – to consult on the best methods, the best sort of jar to keep your samples in, the best place to find the rarest species.

This fashionable mania had vast consequences. The enthusiastic naturalists of the late 18th and early 19th centuries – a few of them professionals connected to museums and gardens, many simply amateurs, curious locals or travellers with a practical fascination for the places they went and the things they saw – were actually laying the groundwork for one of the greatest revolutions in our concept of the Universe: the theory of evolution. Without the vast effort of mapping the network of species covering the planet, the fundamental pattern of evolution could never have been glimpsed. We would have no genetics, no DNA – essentially, no science of biology.

The making of a botanist

Arriving in Edinburgh in 1790, Robert Brown attended lectures on natural history as well as touring around collecting plants in such places as Leith shore and Musselburgh just along the coast. In the summer holidays he took excursions into the Highlands with friends, again collecting plants. He hung around the fledgling botanical garden on Edinburgh's Leith Walk, making friends with his fellow *flora* enthusiasts. He walked in the Scottish hills whenever he could, accompanied on some occasions by a young fellow-student called Mungo Park. Park became one of Scotland's most celebrated explorers, striking into parts of Africa that no European had ever seen – an example Robert matched in that other mysterious continent, Australia. Robert scoured the hills around Edinburgh for unknown species, strange mosses and heathers, bringing samples back to the Leith Walk garden. At 18 he presented a paper to the Edinburgh Natural History Society, pointing out errors in the standard text of Scottish flora of the time. In 1793 he began compiling a book of plant descriptions. Its meticulous attention to detail prefigured his later careful experiments and scientific work.

Though he continued his medical studies, Robert's heart was not in it, and he never did graduate from the University. In today's world, where qualifications and exam grades rule, Robert Brown could never have been a practising scientist. But whatever he lacked in official university certificates, he gained far more in experience. Those youthful wanderings around the Scottish hills gave him a deep appreciation of the structure of the living world. In 1793 Robert discontinued his university studies, and in October of 1794 he was commissioned Ensign in the new Fifeshire Fencibles regiment, based at Cupar, a small town not far from St Andrews.

Robert soon found himself posted with the regiment to Ireland. The commission was an almost ideal appointment for him. Despite the growing conflict between the ruling Protestant English and the Catholic Irish, the Fifeshire Fencibles saw little action, and Robert found ample time to continue collecting in a land unfamiliar, a place full of new botanical surprises. By May 1795 even his incomplete medical studies proved useful: he was appointed Surgeon's Mate in the regiment. Brown soon familiarized himself with the botanical facilities and personalities available to him in Ireland. His letters to various fellow botanical enthusiasts abound with debates and disputes on identifications, careful descriptions of plants, and enquiries after the latest publications. Nevertheless, he complained from time to time of the lack of organization of botany in the locality: there were, for instance, few established 'herbariums' where he could compare his findings with known, named specimens. In one letter he wrote that the conditions of his life at that time 'will never allow me to be more than a smatterer in the Science'. He could not have been more wrong.

Army life wasn't all botanizing. Brown took occasional and apparently lacklustre part in military exercises. He had his medical duties, such as treating the gunshot wounds of a friend injured in a duel, dealing with the venereal ulcers of various soldiers, and even pretending at one time to try to revive a man clearly drowned, 'merely to satisfy the bystanders'. There was also the regiment's social life. In his diary Brown was meticulous in recording the quantities he drank: from pints of wine to bottles of port, from gin punch to cider. The drinking, in fact, combined with a backgammon habit, began to cost him dear. He played the violin and taught himself German. There are more meticulous diary entries on his own medical symptoms – foreshadowing a somewhat hypochondriac tendency in his later years. Eventually army routine began to seem tedious and without prospect of change or advancement, and Robert began to cast around for a way out.

In 1798, on a visit to London, he managed to obtain an introduction to scientific grandee Sir Joseph Banks[1]. Robert's reasons were entirely practical: he wanted to study the Scottish samples in Banks's collection, at the time the finest in the country. Sir Joseph Banks was a prominent man, with dealings across the whole range of sciences, a member of many influential clubs and societies, and a friend of the King. For the past twenty years he had also been President of the Royal Society, the grand body of British scientists. The unknown 26-year-old Robert Brown had to go through two intermediaries to obtain access to Banks's collections in his Soho Square house.

Joseph Banks was, though, a very good contact to have, and well worth the effort, beyond the immediate advantages of access to his botanical collection. Robert Brown's first visit to Banks's house was the beginning of what would be a very long association between the two men.

Joseph Banks had made his own name as Ship's Naturalist on Captain Cook's first voyage to Australia, thirty years before; it was Banks's collecting activities that inspired Cook to name the famous Botany Bay. And it was Banks who eventually sent Robert Brown in his own footsteps, out to the new continent again. Australia – actually still called 'New Holland' at the time – was where the young Robert Brown would really cut his scientific teeth.

---

1 While in the capital Brown was able to go plant collecting, gathering samples from ditches around Pimlico and Battersea and the banks of the Thames.

Journey to a New World

The year 1801 was not an auspicious time to find oneself riding the ocean wave. Britain and France were engaged in a desperate struggle for maritime supremacy, as Napoleon Bonaparte rampaged across Europe. The one place where Napoleon's apparently insuperable tactical skill might be challenged was upon the sea; and such a challenge fell naturally to the island nation of Great Britain.

War or no war, scientists still wanted to do science. Both French and British ships continued to ply the waves in search of new worlds, new plants and new creatures. Sir Joseph Banks had for some time been nursing a plan for a voyage to explore the interior of New South Wales, believing the region ripe for the exploitation of natural resources. From a hard-pressed Navy, Banks managed to secure a rather leaky old ship called the *Investigator*, and a certain Captain Flinders to sail her. But no exploratory voyage would be complete without a resident naturalist.

Robert Brown's inclusion on the trip to Australia was more or less accidental. By 1801 Banks himself was too old to go gallivanting around the world. He had previously employed Brown's old Edinburgh chum Mungo Park on voyages of exploration to Sumatra and to the Gambia[2]. Park turned down the poorly paid post of naturalist on the new Australian venture: he was getting married soon, to a surgeon's daughter from Selkirk ('a love affair in Scotland but no money in it', as a mutual contact described it to Banks) and he needed all the finances he could get.

After Mungo Park's refusal, the young Robert Brown came well recommended: 'He is a Scotchman, fit to pursue an object with constance and cold mind', said a friend of Robert's whom Banks turned to for advice. Despite delays, rearrangements and political manoeuvrings, Banks finally wrote to Brown in December 1800

---

2 On both those voyages Park was reported perished, only to turn up again unexpectedly. He finally did really perish in Africa, disappearing in 1806 during a pioneering trip up the Congo river.

offering him the post of naturalist on what had mutated into a coastal surveying mission, with plans to circumnavigate the entire continent of New Holland. At the beginning of the 19th century it wasn't even known for certain that the landmass was a single continent: some suspected that it might be a collection of large islands. Brown's salary was £400 a year. The voyage, Banks predicted, would last at least three years.

Robert Brown accepted the offer by return of post. On his 27th birthday, 21 December 1800, Brown boarded the packet boat at Dublin. He crossed the Irish Sea, walked to Chester and caught the coach to London. He reached the capital on Christmas Morning 1800. A new world and a new life awaited him.

Terra Australis

The *Investigator* set sail from Portsmouth at 11 o'clock on the morning of 18 July 1801. In the English Channel they met the first evidence that the sea was a far from peaceful arena in the form of four warships of the Royal Navy, including the *Temeraire* later immortalized by the artist J. M. W. Turner. The next day the *Investigator* was fired upon by mysterious boats. The attackers turned tail when a warship appeared on the horizon: they were Guernsey pirates, chancing their arm against the limited defences of Brown's ship. This journey into the unknown was going to be no cake-walk.

The old ship, to add to the problems, was already leaking. The on-board astronomer was incapacitated by seasickness. Happily, Robert Brown, though he had never been to sea before, turned out to have excellent sea legs. Heading south for the Cape of Good Hope, they were soon passing the Azores and Madeira. The scientific contingent, including Brown, the botanical artist Ferdinand Bauer, and Brown's young assistant Peter Good, went enthusiastically about their tasks, eagerly examining whatever creature happened into their path. They dragged an unfamiliar sort of turtle from the sea and dutifully measured its rectal temperature; and they rowed for six hours to one of the tiny Desertas islets, only to find bare rock and nothing of any scientific interest at all. A luminescent swarm of medusae – glowing jellyfish – passed them as they approached the Equator. Captain Flinders was assiduously handing out lime juice and sugar with the grog rations in an attempt to stave off the ultimate sailor's nightmare, scurvy[3]. Crossing the Equator, Flinders dressed up as Neptune in an enactment of a traditional sailors' rite, while the rest of the men got messily drunk. The non-mariners, including Robert Brown, were not too impressed. 'As usual the sailors got drunk and turbulent at night', wrote Peter Good in his diary.

By October the *Investigator* reached Cape Town. Brown and his scientific team climbed Table Mountain – reaching the summit at sunset and nearly coming to grief in the dark and fog on the way down. Brown was in his element, indulging in a plant-collecting frenzy – but compared to their ultimate destination Cape Town was a suburban back garden. So much botany had been done there over the years that there was little new to learn. Nevertheless, years later Brown recalled the days spent on the mountains around the Cape as 'some of the pleasantest botanizing I ever had'.

The *Investigator* set out again as the fleet at Cape Town was firing its guns in celebration of Guy Fawkes day – 5 November 1801. They made good time eastwards, and on 6 December they saw the south west tip of 'New Holland'. Terra Australis at last.

Now the real fun began. Before Robert Brown's voyage only about 400 species of plant from Australia had been properly examined – a tiny percentage of the 12,000 or so unique species that we know of today. Brown ultimately brought back more than 3,000 new species all by himself. In the first three weeks working about the south-west tip of the continent, he collected 500 unknown species. He and Good also saw many animals, including

---

3 It should have been lemon juice, a more efficient protection against scurvy – but medical experts of the time were still confusing the two.

of course kangaroos, but also new species of birds, many insects –
and people: aborigines. On Brown's first meeting with the locals
he had a go at a spot of translating, picking up the words for ears
('twang') and cloak ('wurrit'). Understandably, the aborigines
reacted with bemusement to the European scientists' attempt to
measure them with rulers. Even Captain Flinders was getting in
on the science act, making observations of the effects of magnetic
rocks on his ship's compass – observations that he published on
his return to England.

They headed east along the coast in high summer. The heat
was intense, and the south coast, after excellent beginnings,
proved bare of much botanical interest. Brown suffered from sun-
stroke. The explorers carried out a wholesale slaughter policy
amongst the bird population, partly from Flinders' desire to feed
his men fresh meat and keep the scurvy at bay, and partly because
that was the way collectors did things in those days. See something strange and new – fire a musket at it. At one point 760 'mut-
ton' birds were shot in two and a half hours, while on the aptly
named Kangaroo Island they shot 31 kangaroos, making half a
hundred-weight of heads and tails to go into a giant stew.

Even so far from home, the current European conflict wasn't
forgotten: meeting a French vessel, Flinders prepared the ship for
military action. Fortunately, it turned out to be Le Géographe, the
French equivalent of the Investigator – more loaded with botanists
and plant samples than gunpowder and cannons. Technically by
then the two countries weren't even at war: a peace treaty had
been signed in March 1802. But that news hadn't quite reached
New Holland yet.

By May 1802 the expedition was at Port Jackson, near Sydney:
the centre of the British presence in Australia. After a couple of
months spent patching up the ship, botanizing and revictualling –
1483 gallons of rum were taken on board, for instance – the Investigator set out once more. This time the goal was to circumnavigate the whole continent – to find out if it really was a single
landmass or not. Existing Dutch charts of the north coast were

150 years old and amounted, according to Captain Flinders, to little more than 'representations of Fairyland'.

By November the ship was in terrible condition: the carpenters estimated that within a few months there wouldn't be single sound timber in her. By February, as they crawled around the coast, the crew were becoming sickly from the bad diet. Flinders, himself suffering from scurvy, decided to run up to Timor in Indonesia in a bid to find new supplies of fruit and meat. With more stores taken on, they then made a dash for Port Jackson, still going anticlockwise around the continent. The crew were falling fast, the bosun and the quartermaster dying from malnutrition. Finally, on 2 June 1803, the leaky, crumbling *Investigator* crawled back into Sydney harbour, having dragged its way around the entire continent of Australia.

Robert Brown had delved into the fantastic wildlife of a new world as no one, not even Joseph Banks, ever had before: 3,600 species of plants, 3,200 of them completely unknown; a hundred or more new insects and animals; and at last a reasonably reliable chart of the entire coast of New Holland. The final stages of the journey reminded them all that such exploration came at a severe cost. Many of Captain Flinders's friends and fellow sailors lost their lives. Peter Good, Brown's trusty assistant, died even as they dropped anchor at Sydney.

Botany was a dangerous business.

Back to the quiet life

Robert Brown spent another year and a half collecting in New Holland[4]. Scientific justice could hardly be done to his myriad findings out in the relatively isolated surroundings of Port Jackson and Sydney. Brown finally returned to England, landing at

---

4   Captain Flinders left Australia before Brown, in late 1803. Brown was fortunate not to have gone with him, since Flinders was interned by the French on the island of Mauritius and didn't get back to England until 1810.

Liverpool after a five month journey, in October 1805[5]. With his artist colleague Ferdinand Bauer – many of whose delicate, precise drawings of Brown's samples have become classics – Brown travelled to London. They reported to the Admiralty on 5 November – to find that there was no one there to talk to them. Eventually they left a note, going off to find lodgings for themselves. Later they discovered the reason for their cool reception at the Admiralty: the Royal Navy had just won the decisive battle of the maritime campaign against Napoleon at Trafalgar: everyone at the Admiralty was busy celebrating, and couldn't be bothered with a couple of almost forgotten travellers.

Things had changed in other ways while Brown had been away. New Holland was no longer the 'vogue' in strange new worlds: partly because of its increasing use as a convenient penal dustbin for undesirables and criminals, in the public eye it had become a rather horrible, nasty place. Politically, the expedition, despite Brown's staggering collection of new marvels, found itself almost ignored on its return.

But amongst the scientists Robert Brown's reputation was made. One expert rated his collection from Australia as 'by far the most excellent that ever resulted from any expedition'. Brown was elected Secretary and Librarian of the Linnean Society in London, the main forum of natural history, and given lodgings in the building they had obtained for their new library – a one-time public house called the Turk's Head, in Gerrard Street, Soho. Joseph Banks counted Brown as one of his closest companions, inviting him to his traditional 'philosophical breakfasts' (Thursdays, men only) and intellectual soirées (Sundays, women tolerated). Brown mixed amongst some of the leading scientific lights, such as the chemist Humphrey Davy, the mathematician John Herschel and the astronomer Hugh Maskelyne. He became

---

5   Brown managed to bring a live wombat back to Liverpool: it lived for two years, becoming quite attached to its owners, liking to curl up on their laps.

known for his peculiar combination of dignified reserve and dry, cutting wit.

It wasn't all socializing. Brown spent the next five years going through his New Holland collections in painstaking detail, examining, describing and classifying. With his Linnean salary he bought himself a new microscope and gradually enlarged his library of reference works. In 1811 Joseph Banks appointed Brown Librarian and Keeper of his collections[6]. Banks's library of books was enormous, and his botanical collection probably the best in the world. From then on Brown spent most of his working time in Banks's library and collection rooms in Soho Square: a place that was the centre of the scientific world of the 1810s. Though Brown's voyaging days were more or less over[7], his scientific work had hardly begun. He settled at the lab bench, in his black coat and blue pantaloons, ready to do some serious work.

The secret of sex – plant-style

Botanical ideas of the 1810s and 1820s were inching toward what Charles Darwin would later crystallize as the Theory of Evolution. Robert Brown noted, for instance, that while families of plants containing large numbers of varieties tended to be spread wide around the world, small families tended to crop up in limited geographical zones. Variety was driven by environment.

Brown saw clearly that nature had constructed not a simple sequence but a complex web of life: in the introduction to his voluminous account of the Australian discoveries he wrote of how organisms were linked 'more after the manner of a network than a chain' – prefiguring some modern ideas about the 'network' structure of complex, evolving systems.

---

6 Banks's previous librarian died in somewhat ignominious circumstances: from complications after an operation for piles.

7 Not completely: he continued to travel about Europe, even, on a trip to Russia, being involved in a traffic accident when his troika crashed.

By 1810 Brown had also begun to realize the specific importance of pollen – the heart of plant reproduction. He had shown
that often one of the surest ways to identify the family or genus of
a plant was by looking at its pollen.

In 1820, at the age of 77, Joseph Banks finally died. He had formally bequeathed his house and collections to Robert Brown, on
condition that after Brown they would be transferred to the British Museum. Brown accepted the post of Keeper of Botanical
Collections at the Museum, as the foundation of a natural history
department. He remained working at Soho Square: he was part of
the furniture there, ensconced amongst the books, samples and
microscopes.

From the beginning Brown was inseparable from his microscope. He had shown how vital a tool it was even for the simple
cataloguer, as accurate descriptions and classifications of species
could often be obtained only by the closest microscopic inspec-
tion. By June 1827, it was time for Robert Brown to go beyond the
catalogue.

In 1827 it was known that the female ovulum in the plant
became the embryo of a new generation – and that pollen was
somehow fundamental. It seemed obvious that the pollen needed
to contact the ovulum to enable fertilization. There were question marks over how pollen particles could physically reach the
ovulum. To some it even seemed impossible. Brown's stated aim
was 'to settle questions respecting the mode of action of the
pollen in the process of impregnation'.

For his first experiments Brown chose *Clarkia pulchella* pollen
grains. This evening primrose relative, native to North America,
was first collected by and named after the pioneering botanist
Lewis Clark. *Clarkia pulchella* has beautiful pink flowers – *pulchella* is Latin for pretty.

Brown's microscope was simplicity itself: just a single lens
attached to a stand, above a sample-stage where the object to be
studied was placed. An adjustable mirror caught the best illumi-
nation, from daylight or candles. The whole thing was solidly

mounted on a heavy mahogany casing to minimize vibration. More advanced microscopes existed, but Brown stuck by his simple instrument. The key to good work was the quality of lenses, for which optical instrument makers of the highest calibre were vital.

That Brown's microscope work was respected throughout the scientific community of the time is illustrated by the visit of the young Charles Darwin, some years later. In 1831, when Darwin was preparing to set out as ship's naturalist on the *Beagle*, he went to Brown as the foremost expert in botanical microscopes, not to mention working in the difficult conditions of a ship-borne collecting trip.

Darwin was impressed: he complimented Brown on the 'minuteness and perfect accuracy of his observations'. He and Brown had other things in common: both had studied medicine at Edinburgh, and neither had graduated with any official scientific qualifications. When Darwin returned from his momentous journey on the *Beagle*, he brought Robert Brown samples of plants from all over the world.

Dancing in the middle world

Brown's notes for 12 June 1827 read: 'The particles have manifest motion... only visible to my lens which magnifies 370 times...'. 'While examining the form of these particles immersed in water', he reports in the paper he published in the *Philosophical Magazine*, 'I observed many of them very evidently in motion...'. These lines turned out to have enormously important consequences in the story of how scientists came to understand matter and the middle world.

Brown's first thought, seeing the wild dance of the pollen, was that he had stumbled on some kind of vital life force. The idea that there might be 'elementary particles' or 'atoms' of organic material, fundamental particles in which the mysterious life force resided, had been around for some decades. But Brown found that even samples of dried pollen, and particles 'steeped in gin for

2 weeks', showed the same incessant motion. He checked out moss spores that were over 100 years old – and again found tiny particles in constant motion.

This could not, he was beginning to conclude, be some 'vital force'. He looked at gums, resins, soot and dust dispersed in water. He ground up tiny pieces of window glass, of minerals and metals, and even pieces of meteorite[8]. Whatever he tried, as long as it could be ground into small enough particles, he saw the same evident motion – the same wild dance as that of his original pollen from *Clarkia pulchella*.

Still he wasn't quite satisfied. Over the next months he did more careful experiments to check for possible experimental causes of his observations, such as evaporation of the water or vibration of the microscope. He convinced himself that the motion he saw was truly inherent to the particles and nothing to do with the external environment. He checked out the effect of particle size: particles 1/8,000 of an inch still danced, though noticeably more slowly than those of 1/10,000 of an inch. He heated up the droplets to high temperature and still saw motion.

Whatever this incessant dancing was, he concluded, it was something fundamental to *matter* rather than *life*. In this material world of tiny particles nothing – alive or dead – could stand still. Everything danced.

Brown's paper 'A Brief Account of Microscopical Observations made in the Months of June, July, August 1827, on the Particles Contained in the Pollen of Plants; and on the General Existence of Active Molecules in Organic and Inorganic Bodies' was only privately printed at first. Brown distributed copies to his friends and colleagues. The scientific establishment soon

---

8　According to his notes Brown also tried scrapings from his teeth, 'having dined the previous evening on roast chicken'. He even ground up a tiny piece of the Sphinx and observed its particles in motion. His position at the British Museum must have been instrumental in helping him get hold of this least likely of specimens.

picked up on it and reprints appeared in the *Philosophical Magazine* and the *Edinburgh New Philosophical Journal*, as well as in translations in Germany and France. There was so much initial confusion that Brown published a second paper, in 1829, to try to clear up some of the misconceptions his first seemed to have generated. A scientific 'urban myth' went around that Brown had indeed found the 'vital elementary particles of life'. Anyone who had actually read Brown's paper could not have thought this: but it seemed that, not for the last time, the public had been scientifically misinformed in the interests of sensation. Michael Faraday, at the Royal Institution, did his best to repair the damage by giving a public lecture on the correct version of Brown's results. Faraday's notes even conclude with the prescient speculation that the phenomenon observed by Brown might have some important 'connection to atomic or molecular philosophy'.

In his second paper Brown defended both the science and the background to his work. Some claimed that his results were already well known, particularly owing to work by a French botanist, Adolphe Brongniart. Brown gave a detailed summary of the existing literature, including Brongniart's findings, and compared it to his own extensive experiments: mounting a convincing case that though not the first to have *seen* the motion, he was the first to have carefully and comprehensively *examined* it. He had studied a whole range of different samples, both organic and inorganic, as well as particles of different sizes; he had done experiments in different conditions – all necessary to concluding that the observed motion was real and inherent to all matter, living or dead.

Ultimately the scientific community sided with Brown. The dancing motion of tiny particles eventually gained the name 'the Brownian movement' or 'Brownian motion'. Brown, though almost accidentally and some years after his death, achieved the ultimate scientific accolade: they named something after him.

Unaffected greatness

Over the next few years Brown travelled widely in continental Europe. Though he demonstrated his dancing particles whenever he got the chance, he never came any closer to explaining the phenomenon.

It was a popular time for touring Europe, so long out of bounds to the British during the Napoleonic period. Scientists were playing their part in restoring international *détente*. Brown began to collect honours all across the continent, being elected a member of the top scientific societies in both France and Germany. Some commentators in England, disillusioned by the lack of British government support for science and British cultural interest in scientists, pointed out that the distinguished botanist Robert Brown was far more esteemed in Europe than in his own country.

The Presidency of the Royal Society had been given not to a scientist but to Duke Augustus Frederick, one of the many sons of King George III, and Robert Brown was seen as one of the few actual scientists in a position of scientific importance, as Keeper of Collections at the British Museum. Brown took a hand in various political machinations of the time, proving instrumental in setting up new institutions such as the Royal Geographical Society. Eventually he was given an honorary doctorate by Oxford University – at the same time as other luminaries such as Michael Faraday and the chemist John Dalton. None of these men had risen through the 'official' channels of education. They had largely taught themselves about the Universe around them.

Brown continued to botanize, to examine, to describe, to collate – ploughing through the still substantial leftovers from his own expedition as well as calling in samples from a host of other explorers. Darwin sent him samples from Tierra del Fuego and the Galapagos, and on his return to England in 1836 brought more goodies for Brown to examine ('For what he wants them', wrote Darwin, 'I have not a guess – if I can summon courage I will ask him, but I stand in great awe of Robertus Brown'). With microscope or not, botany was still often a matter of cataloguing legwork, as scientists tried to form a clearer picture of the family tree of the natural world.

Brown wasn't quite finished with the microscope. In 1831, for instance, while he studied his old bugbear, pollination, in orchids, he made the first detailed observations of the cell nucleus. Others, as Brown stated, had undoubtedly seen that cells contained these dark particles or nuclei. He was the first to realize that it was a general feature of cells from all kinds of plant. This is probably still the work he is most renowned for today amongst biologists and botanists. Brown also observed the transport of fluid along microscopic pathways inside plant tissue, known today as 'cytoplasmic streaming'.

Science does not, typically, make you rich. By 1843, Brown was 70 years old and near bankruptcy. Despite failing health, his finances forced him to continue his work at the British Museum rather than retire. 'He is growing old fast', wrote a visiting American botanist, 'though still full of gossip and... a great deal of dry wit... He knows everything!' A coterie of friends and colleagues managed to persuade Prime Minister Robert Peel to grant Brown a pension from the Civil List: which Brown, characteristically, refused. Instead he lobbied for a pension for the descendants of his old friend Captain Flinders of the *Investigator*, who had died in 1840.

Brown could be 'the driest pump imaginable' -- reserved and silent in company; and yet many visitors found that, got alone by the fireside, he launched into strings of jokes and anecdotes. As late as 1839 he was talking of a botanical visit to America. Instead, in 1840 he visited Germany again for more botanical meetings – taking in Portugal on the way. Meanwhile, in Islington there was a certain Louisa Harris, to whom Brown despatched many letters as he travelled, and who he wrote into his will of 1843. He wasn't finished living yet.

He was still travelling as the middle of the century approached. In 1849 he took a steamboat up to Edinburgh, continuing on to Dundee and his birthplace of Montrose: he saw a collection of his father's sermons and the ruins of the minister's old chapel. He went on to Ireland, visiting the haunts of his distant youth in the Army.

Pleading ill health he declined the invitation to sit on the Commission for the Great Exhibition of 1851: he wrote to a friend 'I am both an invalid and much out of spirits'. He continued to live in Joseph Banks's old house in Soho Square, obtaining a lease from its owners after the last of the botanical samples had been moved out to Kew Gardens and the British Museum. In 1854 the 81-year-old Brown made another, final visit to Montrose, and even climbed to the summit of Lochnagar – a mountaintop where he had gathered plants more than 60 years before.

And *still* he was running the botanical department at the British Museum. At 84 he was showing visitors around the collections as enthusiastically as ever, despite bronchitis, gout, and what he called his 'constitutional indolence and great depression of spirits'. Said an acquaintance, 'He is now feeble in body, but an unaffectedly great man in character'. His Soho Square bookseller described him in the last years: 'A little spare man, always dressed in black... very reserved, and often used to escape into his house if he imagined that a stranger desired to speak to him. This not from an unkindness of spirit, but from a feeling of shyness...'. But still 'the old gleam brightened up his features after some racy anecdote'.

In early June 1858 the doctor was summoned and started a treatment of morphine. A reverend of the church called on Brown and demanded to know if he had thought seriously of death. 'Indeed I have', declared Brown, 'long and often – but I have no apprehensions'. Doctors offered to keep him alive – until Christmas, they said, with the help of opium – but Robert Brown declined, preferring to stay in control of his mind to the last moment.

He died at ten minutes to ten on the morning of Thursday 10 June 1858. By then, the dance of the pollen, Robert Brown's strange discovery of the summer of 1827, was all but forgotten.

Brown's middle world

Brown himself, though fully acknowledged as the greatest of living botanists in his time, soon sank into the lower levels of obscurity. It was 1871, long after Brown was gone, before the term 'Brownian motion' was coined. (In that same year writer and botany enthusiast George Eliot began to publish her novel *Middlemarch*, set in 1831. Eliot has two of her amateur naturalist characters negotiate a deal involving some 'sea-mice' and a copy of 'Robert Brown's new thing – *Microscopical observations of the pollen of plants*'.)

In today's science – biology or physics, popular or academic – Robert Brown rarely gets a mention. (One exception is Bill Bryson's *A Short History of Nearly Everything*: as Bryson discovered, in the science of the 19th century the 'shadowy figure' of Robert Brown keeps popping up no matter how hard you try to keep him out of it.) There is no plaque commemorating Brown's attendance at Edinburgh University; his old house on Soho Square was demolished in the early 1900s to make way for 20th Century Fox; and the house he grew up in in Montrose was also demolished and replaced by a library. For some years a commemorative plaque there saw alternative use as a manhole cover. Brown's gravestone, in Kensal Green cemetery in North London, is crumbling, badly in need of repair. His scientific reputation has not weathered much better.

In biology, Brown's botanical groundwork was overshadowed by Darwin's great synthesis of a few years later. Nineteenth century physicists remained stubbornly blind to the importance of Brown's observations as they plunged into theories of matter and energy. For more than half a century after Brown's dancing-pollen papers were published, while a few scientists dabbled in trying to explain this 'Brownian motion', it was seen as little more than a curiosity: some odd phenomenon that couldn't easily be explained, but that probably didn't mean very much. Brown may have been the first to glimpse into the middle world and report back – but most scientists' eyes were still firmly closed to this strange restless place.

Why should Robert Brown's dancing pollen have been anything more than a curiosity to the scientists of the mid-19th century? Why should people who, after all, had a lot of very exciting stuff to be thinking about, from steam engines to electricity to light waves, have bothered at all with a few pollen particles in a drop of water?

Because the behaviour of Robert Brown's tiny dancing particles was in conflict with what had been the great wisdom of physical science for more than a hundred years. The restless pollen seemed to disagree with Sir Isaac Newton himself.

Even superficially, Brown's observations implied that there had to be room for *randomness* in the behaviour of matter. Any theory that could explain the incessant dance of tiny bits of matter suspended in water had to have the randomness of that dance built into it. Somewhere in the rules for how matter behaved, it must be possible to escape the straitjacket of precise, determined, lawful motion – and dance.

But matter was supposed to follow fixed regular rules – moving according to sensible, calculable forces. These rules had been laid down – and proved against observations of planets, for instance – by none other than the great Isaac Newton himself, back around the end of the 17th century. There was no room for randomness in Newton's simple rules. If you believed Brown's results at all, then you had to admit that something was going seriously wrong.

Or was it Brown who had gone wrong?

If he hadn't been such a persevering and conscientious observer, it might have been possible to simply dismiss Brown's observations. That is indeed what some scientists tried to do over the next half-century. In the end it wasn't Brown who had gone wrong – he had worked hard to make sure of what he saw, a fact reflected in the detailed papers he wrote describing his experiments. No, the mistake lay with Isaac Newton.

It was a mistake that would take a long time, even after Brown, to put right again.

To see how all that came about, we have to go back in time yet again. Back to a time even before the laws of Isaac Newton, a time when randomness and chaos were still masters of the world.

And a few tiny bright lights of reason were about to change everything.

Chapter 3

GARDENS OF DELIGHT, ORCHARDS OF DETERMINISM

Very little is known about the artist Hieronymus Bosch. His date of birth is uncertain. He was an active painter from at least 1486. He died in 1516. In between, he painted. And he painted chaos.

Bosch's works were must-haves for the nobility of his time. But he was no courtly painter, no sophisticated art-world professional. He belonged to a strict religious community called the Brotherhood of Mary, and almost never left his home town of 's-Hertogenbosch, in what is now the Netherlands. Nevertheless, he was internationally famous even in his day. Chaos was popular.

The psychologist Carl Jung, centuries later, found Hieronymus Bosch a spectacularly visual example of the play of the unconscious mind. Bosch was certainly an individual sort of painter. He didn't really fit into the artistic fashions of the times. And yet he seemed, in his paintings, to grasp and portray something deep inside the psyche of the people of the late Middle Ages: a phantasmagorical psychology of fantastic demons and licentious abandon, trapped in a world of human helplessness in the face of magic, superstition and incomprehensible fate – the mediæval mind-set in all its gory colour and senseless anarchy.

Take one of his most famous pictures: the third panel of the triptych *The Garden of Earthly Delights*. It shows a spectacular scene of utter lawlessness and chaos; a nightmare landscape crowded with figures (more than a thousand of them) cavorting in naked abandon. There are insane, senseless combinations of animals and men, chimeric creatures from terrible dreams. It is a beautiful and terrifying painting.

In particular, *The Garden of Earthly Delights* seems to capture a fearful knowledge, a fundamental belief system that it must have been hard for the mediæval mind to counter: the view of the world as a fundamentally senseless, terrible, lawless, murderous place. A place full of chance pleasures; a place equally full of meaningless perversion, pain and death.

A world, in short, that did not make sense.

To Hieronymus Bosch, as to most mediæval Europeans, the only possible logic in all the cruel chaos of the world didn't come from within the visible world at all. To make sense of the world, you had to look to God. God had a plan. If you followed the rules He had set down you might just scrape through into a more sensible afterlife.

Justice would be done in that afterlife. There, and only there, would you get what you deserved. Sin in this world, be damned in the next. Be good now, and later you would be rewarded. Whichever it was, damnation or reward, it had to happen beyond our tangible world, which, as anyone with eyes and ears could see, was nothing but a chaotic senseless mess. Hieronymus Bosch could see it: he painted it over and over again.

Hieronymus Bosch painted right at the end of the Middle Ages. Even in his own time he stood out as something of a peculiarity. He was one of the last messengers from that chaotic mediæval world. By the time of Bosch's death in 1516, that world was on the way out. The randomness of the mediæval world was ready to be replaced – by a world full of rules.

Enter the rulemakers

There is a famous story about the young Italian Galileo Galilei. One day, they say, he was sitting in church, suffering another interminable sermon, glancing idly around in search of distraction, amusement – anything to make the time pass. And his eye settled on a chain, suspended high from the church roof, holding an incense chalice. As the chain slowly swung from side to side incense fumes spread about the church. It wasn't the incense, or even the chalice, that attracted Galileo's attention. It was the chain. The way it was swinging.

Before long the fledgling genius had worked out the rules of the pendulum.

This story may or may not be true. What isn't in doubt is that Galileo did work out and publish an analysis of the pendulum, demonstrating the rules that such a long oscillating chain must obey as it swings back and forth. And Galileo went on, of course, to carry out umpteen even more famous experiments and demonstrations, like rolling heavy balls down slopes to derive the rules of motion and (in the ultimate apocryphal science experiment) hurling stones and feathers off the Tower of Pisa to demonstrate the principle of the acceleration of gravity: that objects fall with a fixed acceleration irrespective of their weight.

Romantic elaboration or not, these stories illustrate an important moment in history. Galileo, like a small number of contemporary revolutionary thinkers, wrought a fundamental change in the way people saw the world. They began to replace mediæval chaos, Middle Ages randomness, with rules. Rules derived, moreover, from direct observations of the way nature actually worked. Science started to get a grip on the way we thought.

What happened to Galileo is well known: struggle with the institutions of religious power, effective house arrest, and so on. He died in 1642. A pinpoint of light went out. Fortunately, by then such formerly rare sparks of rationality were beginning to come together in a weakly glittering network across Europe. A web of ideas was being created by independent-minded people who wanted to understand for themselves how things worked.

In that very same year, 1642, just a few hundred miles to the north and west, another light began to gleam. This light would eventually outshine everyone else, as the Sun does the distant stars. In the year that Galileo Galilei died, Isaac Newton was born.

Laws from Lincolnshire

Isaac Newton, of course, has his own contender for Greatest Apocryphal Anecdote from Science History. Right up there with Galileo's Pisan spectacle ranks that day in the Newton family's Lincolnshire apple orchard, when a falling fruit on the bonce is supposed to have enlightened Newton to the true facts about gravity.

Anecdotes aside, Newton's mathematical construction of the way gravitational forces determined the motion of planets was the most staggeringly adventurous attack so far on the culture of chaos. True, much of Newton's conception of how forces generated motion existed already: what are generally called 'Newton's Laws of Motion' are essentially there already in the work of Galileo. But Newton travelled into space. His aim was to grasp no less than the entire Universe, and place it firmly within reach of the rules. Newton wasn't satisfied with dropping pebbles off the top of a tower. His working data were observations of planets. He assembled the painstaking observations of a generation of skilled astronomers into a mathematical system of the entire Universe.

At the root of Newton's system was the force of gravity. In his theory, gravity worked according to the Galilean rules of force – which state essentially that the effect of applied forces is to cause acceleration, or changes in the velocity of objects. In one sweep the entire mysterious Universe became a comprehensible phenomenon of matter and force: a place where everything followed the rules. The chaos of the mediæval world was banished.

To get a graphic illustration of this larger-than-mankind stroke of genius from Newton, you could do worse than visit the British Library building in London, just around the corner from King's Cross railway terminus. In the courtyard there stands a huge statue, a man-like figure but about four times the size of any normal man: this is Scottish artist Eduardo Paolozzi's conception of the great Isaac Newton. There he is, bent over the world with his measuring callipers – busily measuring mediæval chaos right out of existence.

Newton's mathematical laws initiated a practical revolution in doing science. You could put in the numbers, turn the handle of the equations, and come out with predictions of how such distant and unfathomable objects as the planets moved. Moreover, those predictions were correct. Newton had codified the cosmos in a set of usable rules.

Perhaps even more significant was the philosophical impact. Newton changed the way people saw the Universe. Suddenly it was a realm of matter obeying rules. Given the state of things at a particular moment – the numbers describing where everything was and how everything was moving – plus Newton's laws, you could say exactly what things would be like at any time later. The future, it seemed, was *determined* by the rules.

The Universe was a clock – more machine than mystery.

Over the next century, this idea of a predetermined, entirely calculable Universe, with a predetermined entirely calculable future, became a cornerstone of the philosophy of science. This so-called philosophy of 'determinism' rose to the status of unchallengeable wisdom, as the ghost of Isaac Newton himself rose to the status of unchallengeable oracle.

It was only with the arrival of the Industrial Revolution in the early 19th century, and not entirely coincidentally with the peculiar observations of Robert Brown, that determinism started to falter. Pretty soon the whole mechanism of cogs and gears would be in danger of seizing up altogether.

But before all that came atoms.

On the nature of stuff

For many thinkers of the revolutionary 17th century, there was a problem easily as interesting as planetary motion. Just what is stuff made out of?

By the middle of the 17th century scientists had shown that even the apparently empty space around us consisted of something. Champions of the new 'experimental' natural philosophy such as Robert Boyle in Oxford experimented with bell jars and pumps, showing that 'invisible air' could be removed from sealed vessels. You could suck *something* out of space and leave a space far emptier, a vacuum.

Chemistry pioneers – many of them, including Isaac Newton, actually searching for the arcane secrets of alchemical transmutation – had shown that whatever matter is, it tends to come in a few well-established flavours. There are metals, like gold, lead, iron and copper. There are other peculiar substances, like sulphur, phosphorus, air and water. There's more complex stuff, such as blood and urine (many of the alchemists had a peculiar obsession with urine). Many substances, the early chemists concluded, combined with other substances in reliable ways, quite as if there were some fixed set of fundamental recipes that determined what stuff you could make out of other stuff.

Chemistry, confused mix of magic and science though it was at that time, implied then that matter came in some basic building blocks. The basic building blocks had fixed characteristics that led them to combine in predetermined, invariable ways: just as a chef will tell you what you need to make a white sauce or a sponge cake.

What were these basic building blocks? Enlightenment scientists and philosophers shared one assumption. Robert Boyle on the heating of iron nails by the blacksmith's hammer: 'The hammer... impresses a vehement agitation of the *small parts* of the iron...'. Robert Hooke, similarly on heat and matter: 'Heat being nothing else but a very brisk and vehement agitation of *the parts* of a body...'. Francis Bacon, before both Boyle and Hooke: '... and the struggle in *the particles of the body* is not sluggish, but hurried and with violence...'. The philosopher John Locke talked of 'the *insensible parts* of the object'. The Swiss mathematician Daniel Bernoulli, in the early 1700s, constructed a mathematical theory of gases based on the idea that a gas was composed of many tiny particles.

There was consensus, in other words, that stuff was made from many tiny pieces.

No one had ever seen these tiny pieces. What were some of the most rigorous thinkers and experimenters in history doing accepting a theory of matter based on invisible particles they had never seen?

They were building on thousands of years of tradition. The idea of matter being made of tiny particles had been rattling around, in philosophical form at least, since the distant days of the Ancient Greeks. It was an unproven hypothesis that was already 2,000 years old. It was the atom.

The Epicurean atom

The first surviving references to the Ancient Greek idea of the atom come from the works of two philosophers, Leucippus and his pupil Democritus. We owe the first comprehensive theory of matter, and its place in the Universe, to a third philosopher: Epicurus.

Epicurus was born in 341 BC, on the Greek island of Samos. He was drawn to philosophy as a teenager: told of the theory that the Universe had somehow emerged from primeval Chaos, he found himself puzzled. How had this amazing event occurred? He was troubled to find that his teachers were no less puzzled. By the 320s BC he was on military service in Athens, then the centre of the philosophical world, where he may have heard lectures by the great Aristotle. After ten years' further training, Epicurus established his own group of like-minded thinkers in Athens in 306 BC.

At the heart of his philosophy was not, as the detractors claimed (and still do), the desperate pursuit of meaningless pleasure. Epicurus actually recommended striving for a state of serenity where those desires no longer operated, where all needs were banished. Even the fear of death would no longer apply. In such a state there would be time and freedom to contemplate life, to understand it beyond its immediate physical drives. This was the only way to truly feel oneself alive: and death, after such a satisfied life, could hold no threat.

The connection between Epicurus's state of serenity and atomic physics may not seem obvious. His philosophy of spirit stems directly from the question of what stuff is made out of: his physical, material theory of the Universe. Developing the ideas of his predecessors Leucippus and Democritus, he argued that the Universe was basically made from just two things: matter and space.

Matter made up all the things you could touch, see, smell and experience. All these things – water, air, wood, flesh, stone, sand etc. – only *appeared* different, he argued. In reality, matter came in the form of combinations of identical basic units called *atoms*. 'Atomos' in Ancient Greek means 'indivisible': atoms were the bottom line. These atoms could combine in fantastically varied ways to make the fantastically varied stuff we see all around us.

And then there was space. If the atoms were going to combine to make all the stuff around us, they needed space to move around in, so they could meet and form those combinations.

The important point here was that *the atoms were moving about*. This idea, that matter was made from tiny pieces that moved, much later formed the basis of our modern theories of matter. It remained just an idea for more than two thousand years – until the middle world offered up the key to verifying it.

Epicurus's physical theory of matter fed into his advice that trivial desires and fears should be banished, and that a state of serenity was the best target in life. If we are just particles of matter combining and uncombining, where is the room in that for fear, unhappiness, and the pursuit of ephemeral pleasures?

Just how much his own theories of matter and serenity helped Epicurus we don't know: he suffered from chronic illnesses for most of his life, and died, after a bout of great pain, probably from dysentery, in 270 BC. Though most of his actual writings and teachings have been lost, he managed to leave his mark on history, thanks mostly to a Roman poet who came along a few hundred years later.

The poet was Lucretius, born about 94 BC, author of the epic *De Rerum Natura* ('On the Nature of Things'). In the poem Lucretius gives a complete rundown of the basics of the Epicurean Universe. Lucretius's goal was to convince his readers that an Epicurean life free from fear of the unknown was the inevitable product of knowledge of the rules of nature.

Lucretius has suffered bad press at the hands of sententious critics. The Roman republic was in political chaos, gangs were on the streets, and power plays of bloodletting, torture and assassination were the regular components of civic life. Lucretius's paean to satisfied, quiet, desire-free life went against the Roman principle of 'fight or be crushed'. Epicureanism was seen as a threat to the Roman way of life. The only direct biographical information we have about Lucretius was recorded much later by Saint Jerome. Unfortunately for Lucretius, Jerome had a hatred of 'materialist' ideas like those of Epicurus – dangerously subversive advice such as the possibility that escape from fear and misery was not subject to the power of God but actually attainable by anyone who believed in an ordered world of atoms. Jerome was not likely to give a very balanced account of Lucretius. He claimed that Lucretius had been driven mad by a love potion, wrote (under the influence of his insanity) incompetent verse that had to be rewritten by proper poets, and finally committed suicide.

The Orchard of Determinism

So much for ancient philosophies of the atom. By the time of Isaac Newton, even though it was based on essentially no evidence, this concept of matter as many tiny indivisible atoms had been more or less accepted. If indeed a lucky apple did happen to fall on Newton's head that apocryphal day in the Lincolnshire orchard, then taken with the idea of the atom, it signified the beginnings of a new philosophy.

If all matter obeyed Newton's rules, and all matter was made from atoms, then the behaviour of everything, from planets to people to atoms, had to obey the rules in exactly the same way.

Atoms are predictable, everything is made of atoms, *ergo* everything is predictable. So you think you decide when to speak, when to turn your head, or even when to think a particular thought? Think again. According to the Newtonian philosophy of determinism, if you and I are made of atoms then those atoms are behaving entirely predictably, simply obeying Newton's laws.

Imagine that some scientific fortune teller was present at your birth, measuring and recording the state of every piece of matter comprising baby you: the position and motion of all your atoms at the moment of your birth. From that data, using Newton's laws, the fortune teller could calculate the future of every piece of you. The fortune teller could predict everything you were going to do in life and the moment you would do it; everything you would say; everywhere you would go; every thought you would think. That fortune teller would know everything about you, including, of course, that you are now reading about *him* in this book.

It sounds frightening. It also sounds absurd.

When you were born your body consisted of about 200 million billion billion atoms (let's not even worry about the few hundred million billion billion more that you've collected since then). Our fortune teller would have to simultaneously measure the positions and motions of every atom, then calculate their future motions according to Newton's equations. No computing machine will ever be able to do this, not to mention do it quickly enough to actually make a prediction. Though deterministic Newtonianism demands the scientific fortune-teller's predictive ability to be true in principle, in practice such predictive power is utterly impossible.

The arguments about such extreme determinism continued long after Newton's laws themselves were accepted as correct. Though it sometimes led into apparent practical absurdities, determinism as a principle was powerful stuff. It reached its peak toward the end of the 18th century in the work of the French mathematician Pierre-Simon de Laplace.

Laplace was born in Normandy, France, in 1749. As a mathematician he stood head and shoulders above the rest of his arcane profession, reaching the pinnacle of his powers and political influence just as France reached its greatest crisis for many centuries: the Revolution.

It was a dangerous time for thinkers and scientists. But Laplace knew better than most how to bend with the prevailing wind. Before the worst of the Revolution he wisely left Paris, to return only when the Terror subsided and Napoleon Bonaparte took power. Napoleon consulted Laplace on the new Revolutionary Calendar. Though he knew very well that it was mathematical nonsense – the calendar simply didn't match astronomical data – Laplace gave it the OK.

As well as being a clever political operator, Laplace was also one of the greatest mathematicians the world had ever seen. He saw behind Newton's laws the secret of the Universe: that all – future, present and past – could be calculated. With Laplace's weight behind it Newtonian determinism became an incontestable basis for all scientific thought about the behaviour of matter.

Ironically, though Laplace did so much to establish determinism as *the* theory of the Universe, he actually did it to demonstrate how absurd it was from a practical point of view. Because, Laplace pointed out, in almost every real situation there were so many separate influences that useful calculations based on deterministic laws were impossible. The calculations would be immense and practically unmanageable.

Laplace argued that you needed a practical way to deal with the vast swathe of data that was inevitably involved in any real world problem. You needed a way to broad-brush the data: you needed statistics.

Laplace was a pioneer of the mathematics of statistics. Unfortunately for him, he has tended to be better remembered as the champion of extreme determinism rather than champion of statistics. Unfortunate because, as the 19th century progressed, it turned out that Laplace's real love, statistics, was just what scientists needed to understand the mistake they had made in their big leap from planets to atoms.

Rules rule OK. OK?

Combined with Isaac Newton's masterly reduction of the motion of the planets to simple mathematical rules, the idea that matter was made of atoms was very convenient indeed. It meant that Newton's rules could be taken over, lock, stock and barrel, from the big world of planets to the microscopic world of atoms. All matter, on all scales, obeyed the same simple rules.

Downsizing from objects the size of planets and cannonballs to atoms was where the Newtonians tripped up. Leaping straight from planets to atoms, it was assumed that Newton's laws must describe everything in between. Only when scientists began to visit the middle world was it slowly understood that matter was really capable of a lot more than first seemed obvious from Newton's laws.

But immediately post-Newton, post-Laplace, as the 19th century dawned, it sounded as if everything was more or less settled. With the Sword of Determinism all phenomena could be predicted and explained. Newton's simple rules took care of every possible activity any piece of matter might ever indulge in.

Which was one reason why Robert Brown, and other scientists who read about his observations, were flabbergasted by the random, incessant dance of Brown's small particles. This was something to do with matter, and as such it ought, according to the prevailing wisdom, be explicable via Newton's simple laws, where randomness was outlawed.

There was another even deeper riddle. How could the random dance continue so incessantly? In the Newtonian world, when forces were exerted they caused accelerations – changes in speed or direction. In the absence of some other driving force, the pollen particles ought gradually to succumb to the constant friction of the water around them, and slow down and stop.

Had Newton got it wrong? That was a controversial conclusion to draw from a few specks of dust jiggling in a drop of water, no matter how carefully Robert Brown had recorded his observations. Newtonianism represented the great triumph of reason over mediæval chaos: there were laws to the way matter behaved. There was a plethora of hard evidence that Newton couldn't be wrong: the laws worked too well, they enabled calculations of everything from how a planet orbited the Sun to how a cannonball hurtled across a battlefield.

So, while some scientists were surprised by Robert Brown's observations, they weren't surprised enough, in the first half of the 19th century, to risk shaking the Newtonian foundations too vigorously. On its own, Brownian motion wasn't yet enough to challenge the bedrock of Newtonian science.

Instead, something else came along to challenge Newton. To borrow an illustration from modern playwright Tom Stoppard in his play *Arcadia* – there was a problem with rice pudding.

Chapter 4

FROM INDUSTRIAL REVOLUTION... TO RICE PUDDING

The grand question of the industrial age was how best to turn energy – from fires, rivers, the wind – into usable power. What were the *rules*? From this question the whole new science of thermodynamics was born. From its practically minded origins in the hands of gruff no-nonsense engineers, thermodynamics became a triumph of 19th century physical science.

Yet, in figuring out the rules of energy, scientists accidentally illuminated just how these rules disagreed fundamentally with the rules of matter they thought they already understood. Thermodynamics put a spanner in the works of the great Isaac Newton.

English engines and French fundamentals

By the early years of the 19th century the industrial engine, in one form or another, was becoming ubiquitous – perhaps nowhere more so than in Great Britain. Thomas Newcomen, a mining engineer working in Cornwall, constructed the first successful steam-driven pumping engine to remove water from tin mines as far back as 1712. Water-driven mills dotted the countryside, powering the huge expansion of the textile industry. Through the 18th century a series of talented engineers, chief among them James Watt, added step-by-step improvements to the steam engine until it became the engine *de choix* for the industrialist.

Despite all this engineering genius, still there was no fundamental scientific understanding of the principle of the engine:

what rules governed the conversion of energy into useful work? The practically inclined British engineers might have been perfectly happy to live with this gap in the science of energy. On the other side of the Channel was one Frenchman who was not.

Sadi Carnot was a soldier, engineer and fledgling intellectual. Born in 1796, the year after the Revolution, Sadi's early years were lived at the heart of a France in turmoil as the Terror gave way to the rise of the Emperor Napoleon. His father, Lazare Carnot, served as Napoleon's minister of war between 1799 and 1807. Sadi attended the Ecole Polytechnique in Paris to be taught by some of the leading scientific lights of France, such as Poisson and Ampère. Following in the family's army tradition, Sadi went to the military engineering school at Metz in north-eastern France.

Frustrated by the petty jobs he was fobbed off with in the army, Sadi returned to Paris in 1815 and began to get interested in France's industrial and economic problems. He showed a fierce curiosity about the world around him, attending university courses, touring factories and workshops, and making extensive notes on economic and political theories. In 1821 he visited his father in Magdeburg, and suddenly his curiosity found a new and definite focus.

A science of engines

A new steam engine had recently been constructed at Magdeburg, and the Carnots – Lazare, Sadi and Sadi's brother Hippolyte – were inspired into lengthy discussions about how to make steam engines work better. They were a family where engineering ran in the blood. Sadi was particularly frustrated by the backward state of affairs in France, which, during the Napoleonic wars, had been excluded from the many advances made in the rest of Europe, especially in enemy Britain. This, Sadi saw, was the key factor that would determine the successful rebuilding of France as a great power: mastery not just of the technology but of the science of engines.

Britain, at that time, was leading the charge into the industrial future. Sadi Carnot noticed that most of Britain's technological advances had been made by gifted engineers, not formally trained scientists. Rather than worry about the underlying principles of a problem, they arrived – usually by trial and error and a lot of elbow grease – at clever practical solutions. Britain's head start in the Industrial Revolution came from tinkering and ingenuity rather than sound scientific principles.

Sadi returned from Magdeburg in 1821 with a mission: to work out a *theory* of the steam engine. To find the principles that underlay how much work you could get out of it per kilo of coal you put in. Perhaps the French could still steal a march on their old rivals, the unscientific British.

The British engineers such as James Watt had played with the design of the steam engine, trying essentially to answer the same question in a trial and error way: what happens if we change the temperatures of the vessels, what happens if we change the pressure, what if we use some other substance instead of steam? Enormous gains had been made, but, as Carnot realized, the British engineers were rather like opportunistic thieves stumbling around in a treasure-house in the dark. Occasionally they tripped up over something very valuable and were very pleased with themselves. But until someone – a clever scientific Frenchman for instance – found the light-switch, the poor ignorant Brits would never know what vastly more fantastic riches they'd been missing.

Carnot's insight, gained by considering the various examples of engines in use in the industries of the time, was that to make use of energy you needed some kind of *difference*. A water-driven mill, for instance, depended on a height difference across which water fell to drive a wheel: deriving work from the gravity of the Earth. In a steam engine the difference was one of temperature, between a hot fire burning fuel to make steam and a cold condensing vessel where the steam was turned back into water, creating a vacuum into which the pressure of the atmosphere drove the engine's piston.

All engines, Carnot saw, worked between two levels. They all depended on two 'reservoirs' of energy. The difference between these sources enabled a flow of energy between them, and the conversion of some of that energy into useful work.

Carnot asked another perceptive question. He imagined a perfect engine, producing work at the maximum efficiency, by converting heat to work as it flowed from a high- to a low-temperature reservoir. What determined how good that perfect engine could get? What did the maximum efficiency – the maximum amount of work out per unit of energy put in – depend on?

That theoretical maximum was determined, Carnot realized, simply by the temperatures of the two reservoirs. Carnot showed that it could not depend on the details, for instance, of the substance used in the engine (i.e. steam or some other gas). There was a fundamental limit that no engine could surpass, no matter how clever the engineering.

Carnot's reasoning was based on an observable fact of reality: that heat never flows from cold to hot. If such a spontaneous 'cold to hot' heat flow were possible, you wouldn't need to plug your refrigerator into the power socket, as heat would naturally flow out of it and cool down its contents without any external power. Of course this doesn't happen. Heat only flows one way.

Carnot showed that if the maximum engine efficiency depended on the details of the engine – if it could be improved without fundamental limit by tinkering with the engineering – then that would allow a more efficient engine to be used to drive a less efficient one in reverse, and actually achieve a net flow of heat, without any input of power, from the cold to the hot reservoir, thereby violating a principle of nature that was plain for anyone to see.

Engines are not limited only by engineering, Carnot realized: ultimately they are limited by the fundamental nature of energy and heat. Suddenly this wasn't just a theory of engines, it was a theory of nature. Carnot had taken the first steps in turning centuries of speculation about heat and energy into a respectable scientific theory.

Carnot's perceptive vision of the fundamentals of the engine, and how they related to the science of heat, ought to have guaranteed him fame and fortune. Industrialists whose success depended directly on the efficiency of their engines made vast profits. Surely Sadi Carnot was due a slice of the bounty? Things didn't quite work out that way. In the 1820s Carnot was still ahead of his time: his work fell on deaf ears.

In 1824 Carnot published the sum of his musings about heat, energy and engines in a book titled *Réflexions sur la Puissance Motrice du Feu* ('On the Motive Power of Fire'). The book didn't exactly fly off the shelves. It went almost completely unappreciated. Unfortunately, Carnot didn't get much time to champion his theory. In 1830 he caught cholera. Weakened by a previous illness, he succumbed quickly, and was dead within days.

One-way heat and a dying Universe

Some years later another young scientist, on a visit to a Paris laboratory to conduct some experiments on thermometers, happened across an old copy of Carnot's book in a second-hand bookshop. The scientist was William Thomson, better known later as Lord Kelvin.

William Thomson was born in Belfast, Northern Ireland, in 1824, the same year that Sadi Carnot published *Réflexions sur la Puissance Motrice du Feu*. By age seven, Thomson, now moved across to Scotland, was attending his father James's lectures on mathematics at the University of Glasgow. He enrolled as an undergraduate there at the age of 10; published his first original paper at 15; and was studying mathematics in Cambridge, when he could fit it in between rowing races, by 17. It was just after graduating from Cambridge that, on the advice of his father, he went to Paris to learn the trade of the experimental physicist. Thomson's father had a plan. He knew that, back in Glasgow, the incumbent Professor of Natural Philosophy (i.e. physics) wasn't getting any younger. When, in 1846, that professor finally died, Thomson senior triumphantly shoehorned his 22-year-old son into the post.

Not that William didn't deserve it. At age 34 he was knighted for his contributions to the first successful cross-Atlantic electrical communications by ocean-floor cable: he designed ultra-sensitive instruments to pick up the extremely weak signals after their lengthy journey under the Atlantic. Over his life, Thomson made a fortune from various inventions, engineering consultancies and patents: there was nothing of the ivory-towered professor about him. No scientific puzzle daunted him. 'Science', he declared, 'is bound to face fearlessly every problem'. He made substantial contributions to a vast range of disciplines, from electricity and magnetism to elasticity, telegraphy, electrical engineering and navigation.

And, of course, heat. Thomson's first paper, at that tender age of 15, was a mathematical analysis of heat flow. Reading Carnot's forgotten book a few years later put Thomson on the trail of a fundamental consequence of the one-way hot-to-cold flow of heat, going beyond Carnot's reasoning about engines.

Carnot's argument meant, Thomson realized, not only that there was a limit to the efficiency of any engine, but that perfectly efficient processes could never actually be achieved in practice. The efficiency of Carnot's perfect engine depended on the difference between the temperatures of the two energy sources. Only if the low-temperature energy reservoir were at the absolute zero of temperature could perfect efficiency be reached. But the absolute zero itself was fundamentally out of reach.

Less than perfect efficiency meant that at least some portion of the energy flowing from hot to cold must be wasted. Complete conversion of flowing heat to useful work was impossible.

Thomson's great insight came when he realized that this impossibility of perfect efficiency applied not just to man-made engines but to any process: *any* conversion of heat energy to usable work. It applied to chemical reactions, to energy conversion by animals, to the inner workings of stars – to everything.

In other words, all practical or natural processes were going to be fundamentally less than perfectly efficient. Some portion of the heat extracted must always fail to be converted to usable energy, and be lost instead, unused and unrecoverable.

This, Thomson saw, had big consequences – for engineering, yes, but perhaps more importantly for the whole Universe.

Everything that happens in the Universe requires conversion of energy into useful work. But in the process some heat is always being lost. Whether natural process or engineered power production, everything that happens contributes to wasted heat, to a constant, unstoppable deterioration in the amount of useful energy available in the Universe. Everything we do and everything that happens involves squandering some portion of useful energy. It's all downhill. The Universe is doomed.

Thomson called it 'the heat death of the Universe'. Eventually, at some point in the life of the Universe, all useful energy will have been frittered away as waste heat. At that point, no further conversion of energy to work will be possible. That was it. *Finito*. *Fin*. *Kaput*. The End.

William Thomson may have doomed the whole Universe, but notwithstanding this blackest of conclusions he was by all accounts a rather happy fellow. He approached every problem with a boyish simplicity and enthusiasm. He travelled widely, knew everyone and enjoyed everything. He didn't let setbacks get him down: on the death of his first wife Margaret in 1850, for instance, he bought a huge yacht to cheer himself up. Sailing the yacht off Madeira he met his second wife, Frances.

He was refreshingly free from illusions of grandeur in his role as one of the most influential of Victorian scientists. For instance, in 1871 he was refused entry to the British Association meeting in Edinburgh. He'd forgotten his ticket and the porter wouldn't believe who he was. This was a bit of a cock-up, since he was actually President of the Association and was due to give the Presidential Address. But Thomson was merely amused by the whole episode.

Despite his achievements he was known for his scientific modesty. On the occasion of his Jubilee as Professor at Glasgow – he was in the Chair there for 53 years – Thomson declared that he knew no more about electricity and magnetism then than he had done 50 years ago. In 1896 he was ennobled by Queen Victoria, taking the name Lord Kelvin after the Glasgow river. Retiring finally in 1899, he signed himself in the University Rolls, 'Research Student'.

A measure of the one-way world

William Thomson's recognition of the central importance of the one-way flow of heat was the middle step in converting Sadi Carnot's attempt at a theory of engines into a complete theory of energy. Still the understanding of energy lacked something: a way to put numbers into the problem, to quantify heat's irreversible flow.

On the second day of 1822, just about the time when Sadi Carnot was formulating his theory of the engine and just a couple of years before the birth of baby William Thomson, a sixth son was born to the Clausius family of Koslin, Prussia[1]. The head of the family, the local church minister, named the boy Rudolf Julius Emmanuel. Rudolf Clausius grew up to be a pioneer of the application of maths to the problem of heat and energy – not to mention a decorated war hero and famous picker of scientific quarrels to boot.

Clausius's scientific work ran almost in parallel to that of William Thomson. A quarter of a century after Sadi Carnot, the problem of heat finally found its twin champions. It was as if the two men dogged each other's footsteps, both trying to grab the same scientific ground to call their own. They were quite different characters: Rudolf Clausius dented his reputation with fierce academic disputes over who said what first – something that it is difficult to imagine the genial William Thomson being too worried about.

---

1 Now Koszalin, in Poland.

Thomson's and Clausius's scientific parallels weren't confined to heat: Clausius's doctoral thesis, completed in 1847, was on the colour of the sky, which Thomson also worked on. It was as if they were doomed to compete with each other no matter what they chose to think about. William Thomson won the blue-sky contest, his explanation in terms of scattering of sunlight turning out to work better than Clausius's arguments about refraction and reflection. But Clausius can probably claim to be the winner in the heat race.

In 1850 Clausius published a paper in the journal *Annalen der Physik* that finally gave mathematical form to the ideas of Sadi Carnot, distinguishing clearly for the first time between energy and heat. Energy, Clausius realized, is a quantity that comes in various forms, can be converted from one form to another, but whose total never diminishes. Heat, meanwhile, is a special form of energy, distinguished from every other form by its one-sidedness, its unwillingness to spontaneously flow from cold to hot.

With the 1850 paper Rudolf Clausius established the bedrock of the science of thermodynamics: coming up with just two laws to describe the fundamentals of heat, energy and temperature. The First Law was essentially a definition of energy; the Second Law was, at last, a mathematical statement of that principle of one-sided nature: that heat only flows from hot to cold. William Thomson gladly acknowledged that Clausius had won that round of the competition.

So Rudolf Clausius had established the First and Second Laws of thermodynamics. One might imagine that when a scientist publishes a result or an idea – particularly one in the grand-sounding form of a 'Law' – that must mean said scientist comprehends more or less what he or she is saying. This is often – perhaps almost invariably – not the case. If everyone waited until they fully understood everything they would rarely publish a thing. Clausius spent the next two decades straightening out the Second Law of thermodynamics – trying to understand what he himself had come up with.

Those two decades represented an extended bout of scientific thinking that was one of the most momentous, yet somehow one of the most mysterious, in the annals of modern science. The climax came, after numerous further papers and sundry digressions and diversions, in 1865, when Clausius finally gave a name to a strange beast he'd been wrestling with for 15 years: the mathematical animal that personified the one-way world of heat. He called this creature 'entropy'.

Clausius's great contribution was to realize that the concept of 'energy' was not enough to correctly describe processes involving heat and work, such as the expansion or compression of a gas[2]. There was another quantity, Clausius's entropy, that had to be incorporated into any description of such a process in order to comply with the natural principle of the one-way flow of heat.

To describe two quantities you need two laws. Mathematically the First Law deals with energy. It says that energy cannot be created or destroyed, but can only change form. The Second Law deals with entropy. It says that entropy can only increase or stay the same: in any natural process or change[3], because of the one-way flow of heat, the entropy can never decrease.

Sadi Carnot had tried to take account of heat's one-sidedness in his theory of engines. William Thomson had realized that not only engines but every process was fundamentally affected by the one-way flow of heat. Rudolf Clausius, through 15 years of work, set up the mathematical scaffolding of a real usable theory. It put one-way heat flow, in the form of the new quantity entropy, into

---

2 Gases were the favourite stuff of the thermodynamicists; partly perhaps due to the obvious relevance to steam in a steam engine, but also because gases showed the simplest behaviour of substances, much simpler than liquids or solids.

3 Clausius coined the word 'entropy' to recognize the importance of the idea of *change*: 'trope' comes from the Greek for 'change', while Clausius merely took the 'en' from 'energy', to demonstrate that the two quantities, like the two laws, made a sort of royal pair, the two fundamental ruling quantities of thermodynamics.

the heart of how all substances behave. Clausius's Laws of thermodynamics were the fundamental rules of energy and heat. Newton's Laws were the fundamental rules of matter. Post-Clausius, it remained then simply to put the two sets of rules together. A theory of energy and a theory of matter equals a theory of everything: end of story.

However, even as Rudolf Clausius was busily putting the finishing touches to his work, other scientists were becoming uncomfortably aware that these new ideas about energy and entropy did not go together at all well with Newton's mechanics. Something, somewhere, was badly out of joint.

Consider a bowl of rice pudding.

Stirring things apart

Here is an experiment[4] you can do in your own kitchen:

Take one tin of creamed rice pudding (the culinarily advanced may dispense with the tinned stuff and make their own). The pudding may be hot or cold (I suggest hot as it makes eating the results after the experiment that bit more pleasurable). Pour it into a suitable bowl. Next, take one spoonful of jam, of any flavour. Carefully deposit the blob of jam onto the top of the pudding.

Do not stir yet!

Take stock of what you have. A rice pudding. A blob of jam. The two in contact, yet separate. Not mixed. Pudding here, jam there.

Now you can stir. Take a teaspoon, insert into the pudding somewhere near the jam, begin to push the spoon in circles through the pudding. The jam begins to mix – a streak of red (most jams are red, for some reason) pulled into the mix, steadily more and more jam being drawn into the pudding. Continue stirring as long as you like. Eventually, bring the spoon to a halt and

---

4 With acknowledgements to playwright Tom Stoppard and his 1993 play *Arcadia*, where the rice pudding problem is delightfully explored by the young Thomasina and her lecherous old tutor Septimus Hodge.

hold it there, plunged into the now streaky-red or perhaps even pink rice pudding.

Once again take stock. What do we have? A rice pudding. Mixed thoroughly with a spoonful of jam. Pudding here – jam here too.

Now the clever bit. Unstir the jam from the pudding.

How? Obviously, just reverse the direction of the spoon. Follow as closely as you can the path the spoon took as it stirred – but follow it backwards. Retrace the steps of the spoon. Unstir.

See the jam miraculously unmix from the pudding! See the unstirring spoon precisely retrace its steps, drawing the jam back into a nice blob on top of the rice pudding! Stop the spoon precisely where it started. What do you have? An unstirred jam-and-rice pudding! A rice pudding and a blob of jam – in contact, and yet separate. Pudding here, jam there.

Not.

Of course, you don't have pudding here, jam there. Allow me to predict that you still have a pink rice pudding. It has not unstirred. If anything it has just got pinker and even more stirred. Unstirring doesn't work, does it? (If you *have* managed to unstir the jam from the pudding, you have just changed the Universe forever – you have just overturned 150 years of science and more importantly about 15 billion years of reality. Not bad for a quick kitchen experiment. Publish immediately.)

OK, now the technical bit. Unstirring doesn't work. The sequence of 'stirring events' created by the motion of the spoon through the rice pudding proves impossible to 'reverse' simply by turning the spoon in the opposite direction[5]. The stirring of jam into rice pudding is an *irreversible* process.

---

5   Some might argue that if they'd known what to expect they could have been more careful with their unstirring. All right – go ahead. Repeat the experiment – another pudding, another blob of jam. Repeat as often and as carefully as you can. You're only wasting pudding. Once mixed, that's it. There's no going back.

It shouldn't be. A rice pudding oughtn't to be different from any Newtonian lump of stuff. The pudding is made of particles (i.e. rice). According to Newton's Laws of matter, everything that happens in the rice pudding ought to be decomposable into a sequence of collisions between Newtonian particles – grains of rice. Stirring the pudding should simply set in progress a sequence of motions of particles. Rice grains pushed away by the spoon, bumping into each other and bouncing off, moving in response to forces according to Newton's laws.

Such a sequence of collisions between particles ought to be perfectly reversible. Once stirred, simply reverse the initial direction of motion of the spoon, and hey presto! – the same determined sequence of bumps and collisions between grains occurs, exactly in reverse. The result of which should be that, having exactly retraced our Newtonian footsteps, we get back to where we started. Jam *unstirred*. Pudding here, jam there.

Laws that work for planets and cannonballs certainly ought to work for grains of rice in a pudding – unless rice pudding represents some special new form of matter, which doesn't seem likely. Newtonian laws of matter appear to be missing something to account for the obvious irreversibility of the rice pudding problem.

With our little kitchen experiment involving a tin of rice pudding and a spoonful of jam, we have apparently proved that the great Isaac Newton got it wrong. Oops. 'You cannot stir things apart', as Thomasina puts it in Stoppard's *Arcadia*.

The rice pudding and jam experiment is just a nice demonstration (especially nice in that you can eat it afterward). Such irreversibility isn't confined to rice pudding, it is ubiquitous. All natural processes have a one-way sign attached. The ununstirrability, if you will, of stirred jam and rice pudding turns out to be just another example of the principle of one-way nature. Just as heat flows down but not up, so you can stir but you can't unstir.

The rule is the same – the Second Law of thermodynamics. Only processes that increase entropy are allowed. Clausius showed that the flow of heat from cold to hot would mean a decrease in entropy – so it is not allowed. Similarly, unstirring the jam from the rice pudding turns out to result in decreased entropy. Hence the stirring is irreversible.

Exactly how you apply entropy and thermodynamics to a bowl of rice pudding and show that unstirring the jam would indeed mean decreasing entropy isn't immediately obvious. Scientists of the 1860s still had to wait for that insight. But the discord between Newtonian laws of matter and the new laws of heat and energy was laid bare. Newton's laws couldn't deal with ubiquitous natural irreversible processes like the mixing of jam and rice pudding, or indeed (more scientifically) the mixing of different gases. Thermodynamics could.

A mirror crack'd: the disagreement between energy and matter

The two great fundamental jigsaw pieces of all physical science – the science of matter and the science of energy – just did not fit together.

What to do? Throw one of them away? Which one? There were plenty of reasons not to discard thermodynamics. It worked. It was based on half a century of solid observations. And there were plenty of reasons, perhaps even better ones, not to discard Newtonian mechanics. That was based on centuries of hard evidence, and for centuries had shaped the very way scientists thought about nature.

Ditching either of these very successful sciences was therefore not really an option. The only conclusion was that something must be missing.

So it was that in the middle of the 19th century scientists were forced to start thinking very hard indeed about how thermodynamics and Newtonian matter could be fitted together. What was the missing jigsaw piece? And just as it was a vitally necessary line of inquiry, it was also a dangerous one: the attempt to jam the two halves of physical science together brought science even closer to the edge of meltdown. Controversy raged over the very nature of reality itself, not to mention the usefulness of science as a tool to understand it.

Help came from an unexpected source. But not quite yet. First – two geniuses, and some statistics.

Chapter 5
STATISTICS BY GIANTS

Irreversibility was described by entropy. The science of matter had nothing to say about entropy. You couldn't unstir the jam from your rice pudding? The science of matter could only shrug its shoulders and turn sheepishly away.

Something else was needed if the sciences of energy and matter were to be joined properly together. Another revolution: having been led out of the Garden of Earthly Delights by Galileo, Newton and their fellows, scientists of the mid-19th century were now forced to look for a way out of the Orchard of Determinism too.

Around the middle of the 19th century that escape route would finally appear – or rather, be created – thanks largely to just two men. Two über-giants in a scientific century of giants.

The escape route they found was called statistics.

An ugly baby in India Street

In the front room on the ground floor of a classical Georgian terrace in the heart of Edinburgh's New Town, there hangs quite possibly one of the worst paintings of a baby that the world has ever seen.

The baby in this hideous picture grew up and changed the world. Without him, we would quite possibly not have radio, electricity, colour photography, cosmology – or indeed fridge magnets.

The baby was a very young boy named James. The house is part of an imposing Georgian terrace, number 14 India Street. Its lower half is now a modest museum, while the upper half houses, appropriately, a mathematical institute. The baby became the scientist now sometimes called the 'third greatest in history' – beaten into third place only by Albert Einstein and Isaac Newton. He was James Clerk Maxwell.

Maxwell was born at India Street in 1831. He spent just long enough there to be badly painted, before, when James was barely two, the family moved out to their estate in the south-west of Scotland. Aged 10, James returned to the city to attend school at the Edinburgh Academy; and like Robert Brown before him, he even attended the University for three years, from 1847 to 1850, before moving on to Cambridge. At the Edinburgh Academy he was noticed for his precocious mathematical abilities; at the age of only 14 he published his first research paper, on the mathematics of certain sorts of curves. He made friends easily – many he would continue to correspond with for years, and some, such as Peter Guthrie Tait, a future professor at Edinburgh, joined him in the scientific elite of the mid-19th century.

Perhaps it was the Edinburgh air, or perhaps it was just a particularly good time to be alive and curious. Like Robert Brown, James Clerk Maxwell developed a startling desire to probe the mechanisms of the world around him. At Cambridge, he was even rumoured to have studied the well-known adage that cats always fall on their feet – experimentally, by chucking unsuspecting kittens out of his college window. Even as an infant he never stopped demanding to know how everything worked, from the water supply to the bell-wires used to summon the servants.

Through the latter half of the 1850s, still in his twenties, Maxwell busily prepared a sea-change in the way physicists thought about matter. It involved – as thermodynamics always had – thinking about gases, but specifically, thinking about gases as if they were made up of billions of tiny particles zipping to and fro, colliding and rebounding like a furious, microscopic, multiplayer pinball game. Thinking about gases as collections of atoms in other words. This 'particulate' picture of gaseous matter had been slowly evolving since Newton's time, and was undergoing a renaissance just as Maxwell appeared on the scene.

False starts in kinetic theories

The theory of gaseous matter as a microscopic whirlwind of invisible, hurtling, colliding particles went under the name of the 'kinetic' theory of gases – 'kinetic' because, just as in Epicurus's philosophy of matter and space, everything was moving. As far back as 1738, Swiss mathematician Daniel Bernoulli had done calculations of atmospheric pressure that were based on a picture of air as consisting of many moving particles: pressure inside a container of gas was caused by the collision of these particles with the walls of the container. Bernoulli's idea did not, as it turned out, set the world alight. There was no direct evidence, after all, for his 'air particles': it was a mathematical game.

It was 1820 before the kinetic theory was seriously revisited, this time by an obscure British scientist called John Herapath. Herapath's work seemed to be going in interesting directions, such as a calculation of the speed of sound in air that could have been tested against experiment. Unfortunately, various mistakes in the maths let him down. Dismayed by the lack of approval from the scientific establishment, Herapath moved out of science and became editor of a railway magazine. (This may sound like a drift into obscurity, but in 1820s Britain railways were the cutting edge of communications technology: trainspotting was probably a lot sexier than science.)

Next to try his luck with kinetic theory was Scots scientist John James Waterston. Waterston was even more cold-shouldered by the scientific establishment than John Herapath. Born in Edinburgh in 1811, the son of a stationer and sealing-wax manufacturer, Waterston trained as a scientist and engineer, first at the University of Edinburgh and then in London. In 1845, whilst working as a naval engineering instructor in Bombay, he submitted a paper on kinetic theory to the Royal Society in London.

Unfortunately, the Royal Society's esteemed referees rejected it as 'nonsense'. Many nonsense papers were and still are submitted to the Royal Society from the far-flung corners of the crackpot globe; sadly, Waterston's work wasn't nonsense at all. It was, in fact, the best mathematical treatment of the kinetic theory of matter anyone had yet managed: it made key predictions that could have been tested immediately, and would have proved correct. The Royal Society had screwed up.

Half a century later, stumbling across Waterston's never-published paper in the Royal Society archives, the Society's then President Lord Rayleigh admitted the error. Rayleigh estimated that the neglect of Waterston's work had set progress in the kinetic theory of gases back by 10 to 15 years. Not until James Clerk Maxwell and Rudolf Clausius in the 1860s would kinetic theory get back on track.

Waterston himself, returning to his native Edinburgh, continued to more than dabble in science. He published notable papers on topics as varied as the temperature of the Sun, theories of gravity, and the capillarity of liquids. More disputes with various scientific societies ensued: he seemed unable to find any easy way to recognition of his ideas. He became increasingly embittered and withdrawn. Any mention of the learned institutions brought a bitter glare from Waterston. Holed up in his house in Gayfield Square at the east end of Edinburgh's New Town, he became an increasingly reclusive figure.

The end came in 1883. Lucretius the Roman poet was our first suspected suicide amongst the atomists, back in 55 BC. John James Waterston is our second. One blustery morning he set out to walk down to Leith, a northern suburb of the city tucked against the sea. A new breakwater was under construction by the dockyards. It was while walking out along the breakwater that Waterston disappeared. Did he throw himself in, finally overcome by the bitter taste of disappointment and unjustified neglect? Or perhaps a sudden fierce wave took him to a peaceful rest at last, away from the ignorant criticism of a clumsy scientific establishment.

A solution to the problem of stench

Waterston's kinetic theory having been consigned to the limbo of the Royal Society archives, the credit for the first significant mathematical work on the kinetic theory went to entropy pioneer Rudolf Clausius. In the 1850s, ignorant of Waterston's work of a decade before, Clausius started thinking about how the particles or atoms of kinetic theory actually travelled in the gas. Clausius cleared up a difficulty with kinetic theory that can only be called 'the stench problem'.

If indeed it was the travel of particles in a gas that gave rise to the force of pressure – the resisting force you feel when you try to squeeze a balloon, for instance – then it was straightforward to estimate the particle velocity necessary to give forces of the right magnitude. But when these calculations were made, the numbers came out so high that an immediate objection could be raised to the whole particulate theory of gases.

For instance, Manchester scientist James Prescott Joule estimated that an ammonia molecule had to be travelling at about 2,000 feet per second to give rise to the observed pressure. Now, ammonia smells. But when a clumsy chemistry student accidentally spills some ammonia at one end of the laboratory, it takes seconds or even minutes for students at the other end of the class to notice the stench. Given that a typical school chemistry lab is perhaps a hundred feet long, if the ammonia molecules were really travelling at 2,000 feet per second everyone in the lab would smell the spilled chemical immediately.

Clausius pointed out the fallacy in the stench problem. Sure, an ammonia molecule might be travelling at 2,000 feet per second. But it is surrounded by other ammonia molecules. Even in a rather dilute gas, there are a lot of molecules, and they are continually colliding with each other and rebounding off in other directions. Clausius showed that a molecule would travel only a relatively short distance in a straight line before colliding with another molecule and changing direction. In the chemistry lab, vapour from the spilled ammonia would have to edge its way, collision by collision, across the lab: this would seriously delay the transport of smellable quantities of gas, even though each molecule was moving at very high speed between collisions.

'An exercise in mechanics'

Clausius's work, published in February of 1859 in the *Annalen der Physik*, was seized upon by an eager James Clerk Maxwell. He did some quick calculations – 'as an exercise in mechanics' he called it in a letter to colleague and friend George Stokes – as if he were still very much the curious schoolboy, working things out just to see if he could do the sums. Using Clausius's picture of gas particles flying around and colliding with each other, Maxwell tried to predict the gas's viscosity – how it would resist forces such as being squeezed. This was something that could be measured, and what the kinetic theory of gases really needed was some experimental evidence to support it.

Maxwell's colleague Stokes had told him about existing experimental measurements which seemed to show that the greater the density, the greater the gas viscosity. Maxwell was trying to show that these measurements were consistent with the kinetic picture of an atomic gas. But when Maxwell compared his theoretical calculations with the experimental results he was dismayed. His calculations disagreed with the experiment, predicting that the gas's resistance to forces should be the same regardless of its density.

The disagreement seemed to be clear evidence that the kinetic theory of gases was wrong. It did not accord with experimental reality. At first, Maxwell was inclined to accept this conclusion, despite his intuition that the kinetic theory ought to be on the right track.

But he was not quite satisfied. He set about constructing apparatus to do his own measurements of gas viscosity, more careful measurements than those George Stokes had reported to him. He set up a chamber where he could control the gas pressure – and thus the density, from thin vapour to thick pea-souper – while keeping other conditions such as temperature constant. By pushing a stirrer through the candidate gas he could measure its resistance to forces at a wide range of different densities.

What he found, measuring the viscosity of air and other gases, turned his earlier conclusions on their head. His more careful measurements showed that the resistance of a gas is indeed constant, regardless of its density[1]. Thin gas and soupy vapour behave the same. Maxwell found himself in the odd position of proving his own calculation correct, despite having previously declared that it must be wrong. His own experiment had supported the kinetic theory after all.

Maxwell's result, as he belatedly realized, did make sense: though a less dense gas had fewer particles, they moved further between collisions with each other. The two effects balanced each other out, resulting in an unchanged overall resistance.

It was a very unexpected finding, but, for Maxwell, it sealed his belief in the usefulness of the kinetic theory of gases. He was at heart a pragmatist. Though he did not know what atoms really were, and he knew of some facts that the kinetic theory could not properly explain, yet he did know now that the maths worked, and the experiments agreed.

But it was his next insight that set the kinetic theory on the road to real success.

Statistically speaking

James Clerk Maxwell's first thoughts about the emerging science of statistics probably occurred to him at the age of 19, as he read a critical account of a book about the maths of probability in the *Edinburgh Review*. A few years later his burgeoning interest in statistics finally found a focus, when as an undergraduate at Cambridge he decided to have a punt at a particularly nasty problem

---

1   This is true at least for simple gases, those to which the most elementary form of kinetic theory is applicable.

in mechanics. The problem was this: could the planet Saturn really have rings?

Of *course* the planet Saturn has rings. Galileo saw the first evidence of them, and later astronomers confirmed that the gas giant is encircled by a set of rather beautiful coloured discs.

Mathematically the rings of Saturn provoked a bit of a controversy in the mid-19th century. If they were solid discs they couldn't be stable, because the imbalance of forces due to Saturn's gravity ought to tear them apart. The Cambridge University Adams Prize was up for grabs to anyone who could come up with mathematical proof or refutation of the rings' stability, along with an explanation of how come they were or weren't stable.

Maxwell's idea was that the rings might be collections of grains – in the same way that a beach is a collection of sand grains, rather than a single solid object. The problem for Maxwell, in trying to prove the stability of such a ring made from grains, was there were too many of them and no way to calculate all the forces on each. Nor indeed was there any way to know in the first place how big each grain was or how fast it was moving.

What you needed to do, Maxwell showed after three years of effort, was to consider the statistics of the grains. You had to take a democratic view: ask not what every individual grain would do, but rather how the grains behaved on average. By computing statistics, Maxwell found that as long as certain conditions were satisfied, such as that the grains had a certain spread of sizes, then yes, Saturn *could* have stable rings. They weren't going to fly apart. They were OK by mathematics.

Soon after scooping the Adams Prize, Maxwell incorporated this statistical take on grains' motions into his work on the kinetic theory of the molecules in a gas.

Until then most scientists – including Rudolf Clausius, in solving the stench problem – had been happy to assume that all gas molecules had the *same* velocity. It was a simplifying assumption – physicists are fond of such things – and it seemed that no one had thought too much about its consequences. Instead, Maxwell investigated what would happen if the gas molecules had a spread of different velocities – a statistical 'distribution' of speeds. From simple arguments based on probability, he derived a mathematical form for this distribution of speeds. This mathematical function or 'probability distribution' told you just what fraction of the molecules in the gas would have this or that velocity. Many, of course, would be clustered around the average – but there would also be some that were going faster and some slower.

Suddenly the kinetic theory became 'dynamical' – it accepted change, it encompassed variation, it enshrined fluctuation. Molecules could have different velocities. The velocity of molecules also fluctuated over time as they randomly collided with each other. You couldn't hope to predict the velocity of a particular molecule at a particular moment – but you could say what the spread of velocities would be overall. So you could estimate the probability of your chosen molecule having this or that speed.

Like many revolutions, the profound significance of this one wasn't obvious at the time. Pragmatic Maxwell just liked the fact that his statistical method worked. The maths gave a spread of velocities from which all kinds of predictions for the average properties of the gas could be made. As more experiments were done, most of these were vindicated. Maxwell's work at last gave solid support to the atomic theory of matter.

Newton's clockwork Universe was a hopeless place to do real science. James Clerk Maxwell had taken the hint and started doing statistics – replacing real knowledge of individual molecules' motions with averages over statistical distributions. Velocities became numbers that could fluctuate effectively randomly – as long as statistically, on average, as a group, they followed the necessary underlying spread.

The key to just how revolutionary Maxwell's idea was lies in that word 'random'. Newton's rules had no room for randomness, but Maxwell's pragmatism had shown how useful it could be. The chaos of the Garden of Earthly Delights, hidden for so many years as scientists explored the Orchard of Determinism, was glimpsed again.

But what of Rudolf Clausius, who had put Maxwell on the track of kinetic theory in the first place? At the same time as Maxwell was proving the value of statistics, Clausius was becoming embroiled in several disputes over scientific priority. Fiercely patriotic, he saw himself as a champion of Prussian science. In 1870 the Franco-Prussian war broke out – France having been tricked by German federalist supremo Otto von Bismarck into invading the Rhineland – and Clausius extended his patriotism into signing up to fight. He marshalled a team of his Bonn students to act as an ambulance brigade and marched off to the front, earning an Iron Cross and a serious leg wound. After the war Clausius continued to rail over the origins of the theory of energy, refusing to admit that others such as the British had made significant contributions.

This patriotic distraction is perhaps another reason why Clausius failed to follow the latest statistical developments championed by Maxwell. Whatever the cause, Clausius's example of a revolutionary scientist who didn't quite manage to carry his revolution all the way was soon mirrored by another, this time an Austrian, perhaps the most tragic figure in the whole story of the middle world: Ludwig Boltzmann.

Giant #2: Ludwig Boltzmann
If James Clerk Maxwell was a cheery soul, quite the happy Scots laird as well as a mathematical whizz-kid, his partner in the statistical transformation of the science of matter was a stark contrast. He was more the sort who seemed to carry the weight of the world on his shoulders. He was hyper-sensitive, unable to bear the slightest criticism without defending himself fiercely. He struggled with bouts of depression and chronic wanderlust. He never quite settled anywhere; getting caught up in a labyrinth of philosophical problems, he finally found himself left behind by ever-accelerating scientific developments that he himself had started. Where Maxwell wrote humorous ditties, our second statistical superman engaged in dirges about Beethoven and the

afterlife. For him, life was less merry-go-round, more titanic struggle.

This was Ludwig Boltzmann.

Born in Vienna in 1844, Boltzmann grew up in a time of instability across the Austrian Empire and continental Europe. New powers had emerged from the post-Napoleonic, post-Waterloo wastes of Europe in the late 1810s. By the middle of the century these nascent empires were already beginning to clash in the first whispers of a new round of conflict. The empire of Austria–Hungary weathered the gathering storm for half a century – at the cost of falling far behind the bulging powers of Prussia and Great Britain.

For most of the time none of this mattered much to Ludwig Boltzmann. Drawn to physics and mathematics he proved himself a precociously clever and industrious student. Legend has it that the first scientific paper he read was by James Clerk Maxwell; Maxwell and his work became a lifelong guiding light for Boltzmann.

In 1868, at the age of only 24, Ludwig exceeded his *maestro* with a paper on the statistical analysis of the kinetic theory of gases. He showed, using statistics inspired by Maxwell's new approach, that the spread of molecular velocities Maxwell had arrived at by mathematical reasoning alone was also consistent with a more justifiable physical picture of gases, such as gases under gravity in the atmosphere. Essentially Boltzmann generalized Maxwell's work to allow the molecules to have any source of energy – whether simply the energy of their assumed motion, or energy derived from external forces like gravity.

Boltzmann's early career was characterized (as it often is for scientists even nowadays) by a lot of moving around and juggling of job prospects. In 1869 he moved from the University of Vienna to the University of Graz: here, in three tiny unheated rooms, he indulged in experimental work as well as developing his maths. With him were his mother and sister (his father died when Ludwig was only 15, from tuberculosis – after which, said his sister, Ludwig was 'always serious'). As his reputation and his ambition grew, Boltzmann travelled industriously around the top physics departments of Austria and Germany, anxious to make himself known to all the leading lights of the continental physical science world. In 1873, despite having just met in Graz the girl he would eventually marry, Ludwig accepted a post as maths lecturer back in Vienna. It was the undisputed capital of Austrian science and hard to resist, even for someone who didn't really know how to teach maths. In Vienna, Boltzmann continued to play the game, receiving – some would say encouraging – job offers from places like Zurich and Freiburg, only to use them as bargaining chips to raise his status at home.

Meanwhile Boltzmann's insight into kinetic theory grew rapidly. In 1872 – the year Rudolf Clausius spent arguing with the British over who did what first – Boltzmann published a monster paper in the journal *Wiener Berichte*. It was 100 pages of arduous maths, in which he gave the most complete description yet of the statistics of the motion of the gas molecules. It was a *tour de force*. Into one relatively simple equation Boltzmann distilled the essence of kinetic theory: a description of how the statistical spread or distribution of randomly varying particle velocities (or indeed energies) must evolve in response to any situation. Most importantly, Boltzmann showed that, for a gas at equilibrium – settled and steady, not bodily flowing – the form of the speed distribution found by Maxwell was the *only* plausible solution to the equation. Boltzmann's 1872 paper was a firework display of physics, perhaps one of the most triumphant mathematical monographs since Isaac Newton's *Principia Mathematica*.

Unfortunately, nearly everyone agreed that it was also almost incomprehensible.

Entropy without elegance

Did Boltzmann care? Yes – and no. People said his maths was difficult: his explanations came verbatim rather than being calmly considered and carefully arranged. But he had his methods and he was sticking to them. 'Elegance', he famously remarked, 'is for tailors and shoemakers'. After 1872 he knew he had reached a very significant stage in the kinetic theory. From here he could elaborate his ideas knowing that he was going in the right direction, and this he did for the next decade or more.

It was sure-footed, confident, comfortable work. For a brief time Boltzmann seemed to be controlling life rather than it controlling him. In 1876 he won a position as head of the physics department back in his old city of Graz[2]. Vienna might be the centre of the Universe for aspiring physicists, but Boltzmann now had a gravitational power of his own: wherever he went he would be able to make his own centre. Plus, escaping the maths department in Vienna got him out of having to teach any more mathematicians, a task he had quickly grown tired of.

Boltzmann's move to Graz signalled the beginning of his happiest period. Graz gave Boltzmann and his new wife Henriette 14 years of relative peace and contentment, time in which Henriette gave birth to two sons and two daughters and Ludwig indulged his curiosity across a wide range of physics. Despite the inelegant and difficult reputation of Boltzmann's maths, the significance of his triumphant work in kinetic theory slowly began to be appreciated across the scientific world – as a new movement in art, for instance, although initially incomprehensible and even repellent to many, gradually finds acceptance.

Ludwig was far from finished yet. In 1877 he published his next major kinetic theory paper. This time it took the statistics of matter and forced them to the core of the 19th century science of energy, giving the first glimpse of a way to join the two sciences of energy and matter. With the 1877 paper, Boltzmann found the statistical solution to the problem of rice pudding. A mechanical, atomic, *material* interpretation of that mysterious, Newton-violating quantity, entropy.

---

2 The previous head of physics having fallen down a lift shaft – fate works in mysterious ways.

Entropy, Boltzmann showed, was directly equivalent to a rather simple statistical quantity – entropy was a number you could count. The entropy of a sample of gas was equal to the number of different ways in which the gas's total energy could be shared out amongst its constituent particles.

Imagine that 10 schoolchildren have a total of 20 oranges. There are many different ways to share out the oranges amongst the children – just one of them could have all 20, or they could all have two each, or five of them could have three each and the other five one. In Boltzmann's terms, the entropy of this collection of oranges simply counts the total number of different ways in which the 20 oranges can be shared out.

Boltzmann's Great Leap Forward was complete when he showed that his way of calculating entropy also meant that a gas of Newtonian atoms, following reversible mechanical laws, nevertheless could behave irreversibly. It was the solution of the rice pudding paradox. The entropy of the gas, calculated by statistics, obeyed the Second Law of thermodynamics: it could only increase, never spontaneously decrease.

Despite the gas being made of particles following reversible mechanical laws, because it consists of a very large number of such particles its behaviour becomes *statistically* irreversible. Boltzmann showed that in any process, as the constituent particles of the gas move rapidly around colliding with each other, they are bound to share their energy such that the number of possible permutations of the total energy amongst them – the entropy – increases. In the furious chaos of billions of colliding particles, Boltzmann's statistics generate a one-way sign: entropy can only increase.

Boltzmann's entropy formula emerged naturally from the kinetic theory of gases. The kinetic theory had locked into place with irreversible thermodynamics. Boltzmann found irreversibility where everyone least expected it – buried deep inside Newtonian determinism.

Untidy bedrooms and mixed-up rice pudding

As well as opening the way to a kinetic theory explanation of the old Newtonian bugbear, the rice pudding problem, Boltzmann's statistical result also explains another everyday puzzle: why bedrooms get messy so easily.

Consider the untidy bedroom problem: put a lot of hard work in and you can have your room looking immaculate. Now let go – let the room evolve by itself to its natural equilibrium. As every struggling parent knows, a bedroom can become a complete mess in an astoundingly short time – and without more hard work it will stay that way. Why?

It is simply a question of the entropy of tidy and untidy bedrooms. The entropy of tidy bedrooms is low because – following Boltzmann's statistical way of calculating entropy – there is only one way to make a tidy bedroom: every sock, shoe, shirt and pair of trousers in its rightful place.

Conversely, there are many ways to make an untidy bedroom, to arrange its contents in the wrong places. The entropy of untidy bedrooms is high, far higher than that of tidy bedrooms.

Natural processes, such as leaving the bedroom to evolve according to the vicissitudes of everyday life, will always increase the bedroom's entropy. That's the Second Law of thermodynamics. Hence the bedroom naturally departs from the state of tidy – low entropy – toward the state of untidy – high entropy.

It's entirely the fault of the Second Law that bedrooms are such a mess.

Boltzmann's theory and the untidy bedroom analogy also show how entropy is related to the notion of order. Because an 'ordered' (or tidy) state of some system is typically narrowly defined (all the clothes in their correct places) there is usually only a small number of ways of achieving that state. Entropy counts that number of ways; hence ordered states have low entropy. Disordered states, on the other hand, have high entropy: there are a lot of ways of making a mess. Mess is more natural – higher entropy – than order.

But what of rice pudding? By Newtonian thinking, the irreversibility of stirring jam into a rice pudding seems to be impossible. Boltzmann's entropy comes to the rescue again.

The original state of separate rice and jam is an 'ordered' system. Only one possible permutation satisfies it – all the jam in one place and all the rice in another. With only one possible permutation it is a state of very low entropy. Boltzmann's theory shows that such order, once lost (i.e. by beginning to stir), is impossible to relocate: though the particles – this time rice and jam rather than gaseous atoms – might well be perfectly Newtonian, the chaos of collisions always increases the pudding's entropy, so always increases its disorder. Getting back to the ordered, low entropy state of separate jam and rice is impossible.

When it comes to rice pudding Boltzmann's message is this: so what if individually the rice grains are reversible? Statistically, the pudding is not.

A false sense of security

Ludwig Boltzmann appeared to have achieved the ultimate: re-uniting the great themes of physical science, Newtonian mechanics and thermodynamics. Both these theories worked; it was just that they had disagreed with each other. Now Boltzmann had shown that disagreement to be only a matter of statistics. Put enough Newtonian particles together and the irreversibility of thermodynamics emerges as if by magic. The chasm was only on the surface: deep down, the theories interlocked seamlessly.

Perhaps surprisingly, the rest of the scientific community was less than convinced by Boltzmann's triumph. As the last twenty years of the century wore on, Boltzmann found himself at the head of a rapidly dwindling band of enthusiasts for the statistical kinetic theory. It began to seem as if his papers of the 1870s had lulled Boltzmann and the atomists into a false sense of security.

One difficulty was the tortuous mathematics through which Boltzmann derived his results. Even James Clerk Maxwell baulked at the struggle to understand Boltzmann thoroughly, complaining that the more he read, the less he understood. By 1895, twenty years later, British physicists were still trying to get Boltzmann to explain himself properly in the pages of the journal *Nature*. The reception of Ludwig's ideas was even worse in the German-speaking scientific world. By the end of the century many German and Austrian scientists felt that Boltzmann's ideas and the idea of the atom itself had been discredited once and for all.

One of Boltzmann's oldest scientific friends and colleagues was at least partly responsible for the cracks that began to appear in his statistical kinetic theory. This was Josef Loschmidt, a one-time businessman and unfortunately inept entrepreneur who, bankrupt, managed to get a job teaching at the University of Vienna when Boltzmann was first a student, back in the 1860s. Unlike many Boltzmann-bashers, Loschmidt wasn't really trying to knock the kinetic theory: he was himself one of the first to use it to work out realistic estimates for the size of gas molecules. Loschmidt's real aim in questioning Boltzmann's theories was to improve their logic. Loschmidt thought he had spotted a flaw not in the maths, but in the reasoning.

Surely, Josef Loschmidt said, if your kinetic particles obey time-reversible Newtonian laws then there must be at least one way back from disorder to order: in other words it is *possible* for entropy to decrease. Consider starting, for instance, with the rice and jam unmixed (ordered or low entropy). Now stir into a disordered, high entropy state. The Newtonian rice and jam particles have each behaved in a time-reversible way, so if there was a way to get to that stirred state, there *must* be at least one way to get back again. If we *really* reverse precisely every motion of every particle, something that, however difficult to actually do, must be *possible* by Newtonian laws, the stirring would unravel and the pudding would unstir.

Stirred, disordered, high entropy would, in such a case, go to unstirred, ordered, low entropy. Such a process, perfectly possible according to Newton, would lead to a decrease in entropy – perfectly *impossible* according to thermodynamics.

So, said Loschmidt, Boltzmann's Newtonian kinetic gas – and indeed the Newtonian rice pudding – were therefore still not consistent with thermodynamics, because there had to exist at least one way to decrease entropy – hard though it might be to find in a real experiment – and thus at least one way to violate the laws of thermodynamics.

The jigsaw pieces still didn't fit together.

Loschmidt's argument, far from crushing Boltzmann, proved useful to him: it reinforced the fundamental role of statistics. Yes, he replied to Loschmidt, you are correct, there is a way for entropy to decrease. It's not me that's wrong, it's the Second Law of thermodynamics!

We should not, Boltzmann said, be banning entropy decrease – we should simply be realizing how statistically unlikely it is! There are, indeed, one or two ways to decrease the entropy of your rice pudding. But there are billions upon billions of ways to increase it. So unless you are very very very very very very careful to pick just the right way, you will inevitably always see entropy increasing, never decreasing.

Nature will seem to obey the 'Law' that entropy always increases. But only *apparently*: in fact there is no solid Law, only the overwhelming weight of statistics.

Most of us probably feel the same way about the National Lottery. Yes, there is a way to win. But – unless you are very very very very very very lucky – you will *always* lose. It is as if there is a law of nature that says you must lose. Of course there is no such law: it is simply that the chances are stacked too high against you.

Though Boltzmann thought he had won the arguments, many other scientists still did not grasp what he was saying. As the 19th century drew to a close, the disputes and misunderstandings increased in ferocity. Boltzmann began to feel like the last defender of atomism and kinetic theory.

He was probably unlucky in that James Clerk Maxwell, his twin Great Statistical Intellect, had died so young (in 1879). Maxwell's posthumous reputation became nigh on impregnable with the success of his electromagnetic theory (which had, amongst other things, predicted the existence of radio and other electromagnetic waves, subsequently confirmed in experiments). Had Maxwell been there to argue on Boltzmann's side, Ludwig would undoubtedly have had an easier time of it.

In the middle of all this mathematical controversy, what Boltzmann needed to back him up was not more maths but solid experimental evidence: the proof of his statistics in reality, visible for all to see.

Unlikely though it seemed to many around the end of the 19th century, that proof would arrive. But too late for Ludwig.

In the Italian hotel

For many years Ludwig Boltzmann ploughed a lone furrow, defending his ideas from the sceptics and the philosophers. His life did not progress in the ordered manner that might have been expected of a successful academic scientist. In the 1880s, just when he might have been reaping the rewards of his decade of hard work on the kinetic theory, settling himself at the heart of one of the great Austro-German university traditions – Boltzmann somehow lost his way.

He became unsettled at home in Graz. Job offers came in from all quarters, but having accepted one he would change his mind and veer toward another. Wherever the family went – Berlin, Vienna – Boltzmann quickly felt out of place, yearning for yet another move.

Boltzmann's loss of direction in the 1880s also coincided with the death of his mother. Family had always been important to him since his student years in Vienna, looking after his mother and sister as well as playing the precociously clever undergraduate. All through his life Ludwig's work must have drawn him away from his family, first from his sister and mother and later from his wife and children – off again into the minute and complex world of the statistics of atoms. Not to mention the sometimes even more minute and complex world of university politics.

The years passed. Boltzmann began to delve into philosophy – disturbingly spongy ground for a physical scientist more familiar with solid numbers and palpable weights and measures. Perhaps the attacks on his ideas began to wear him down. It was a measure of how profound Boltzmann's ideas were that they drew such deep and complex criticism, and for so long.

Fast forward to 1906. In only a couple of years French chemist Jean Perrin would begin to collect the data that would finally reveal Boltzmann's mysterious statistics for all the world to see with their own eyes.

But Boltzmann couldn't know that. It was in the middle of a family holiday with his wife and daughter, in an Italian hotel room, during an hour or two alone as his family went out for a swim, that he must have come to a decision.

When they returned, his wife and daughter found his body hanging from the wooden window frame.

They took him down, but he was already dead.

An unpopular lottery

Ludwig Boltzmann's embrace of the vagaries of chance proved hard to swallow for many. The lottery of statistical kinetic theory was unpopular.

The laws of thermodynamics – solid, dependable criteria based on a century or more of careful experimentation – were something you could rely on. First Law: Energy was conserved. Second Law: Entropy always increased. These were *Laws* – rules that worked every time. They were not, above all, a question of statistics. There could be no lottery in thermodynamics. Everything must obey.

Ludwig Boltzmann and James Clerk Maxwell put chance and randomness at the heart of the sound dependable rules of how matter and energy worked. No wonder many baulked at such an idea, even if Maxwell's and Boltzmann's calculations gave the right answers.

Most importantly, statistical kinetic theory was based on matter as a collection of tiny particles, all racing around colliding with each other. And no one had ever seen one of these famous molecules or atoms.

Where is the proof of the atom? they said. How do we know these invisible flying entities demanded by your statistical lottery really exist? We need proof that these things are really there before we're going to start taking you seriously.

Even the most convinced atomist had to admit that getting the proof wasn't going to be easy. Actually seeing the atomic entities at the heart of kinetic theory was going to be next to impossible. By Josef Loschmidt's and others' best estimates, if these atoms existed at all they would be far, far too small to be seen with any known method. Besides, they moved too fast. They collided and zinged off in unpredictable directions. A kinetic-theory gas was a furious, unchartable chaos of microscopic motion.

You will never prove it, said the doubters. You will never see these fantastic particles. And what's the use of an atom you can't see – a theory you can't test? It sounded as if no one was ever going to see Maxwell and Boltzmann's statistical world in the flesh.

The doubters were wrong. The reason why was that same ill-advised big leap that the original Newtonians made: jumping straight from planets to atoms, as if there was nothing interesting in between. Since Newton, most scientists had gone on assuming that size didn't matter. They forgot – or never stopped to imagine – that there might be, hidden inside that leap from planets to atoms, a middle world.

That middle world existed. It had, of course, already been stumbled upon, albeit by a botanist rather than a physicist, more than 30 years before Maxwell and Boltzmann. A botanist who had no real chance of grasping the significance of what he had seen.

Finally, as the 19th century gave way to the 20th, it was time for the middle world to divulge a few of its secrets. Time to make the atom real.

Chapter 6
FICTION AND PHILOSOPHY, TIME AND REALITY

The end of the 19th century.

For science, it was – to paraphrase one of the 19th century's literary giants – the best and worst of times.

The way we perceived the Universe around us, and even our own part in it, had changed irrevocably over the previous 100 years. Natural – and technological – processes had been entirely reapprehended, in terms dictated by the laws of thermodynamics: we were in a Universe of energy and entropy, a place where heat, work and temperature were the controlling quantities in the most general theory of reality mankind had ever seen.

The forces of electricity and magnetism had been put together by James Clerk Maxwell – with the same mathematical grace and structure as Newton's theory of gravity a century before.

Chemistry had been transformed from an arcane mystery into a deep and beautiful science of matter, with the Russian Dmitri Mendeleev's periodic table of the elements as the great ordering principle, underlying the chemical behaviour of all substances.

Moreover, all these theories and concepts worked: this was not simply a change in cultural outlook, this was the product of hard experimental evidence. Man had got into touch with reality in a way never seen before.

Of course, there was what many even today see as the greatest scientific idea ever: Charles Darwin's theory of evolution[1]. This

---

1 Even Ludwig Boltzmann, physicist to the core, saw Darwin's theory as the most significant achievement of 19th century science. Evolution

concept added immeasurably to our physical understanding of the Universe. It sought to explain our own place it.

At the same time science was weakening. Its foundations were being gnawed away by a recurring uncertainty. The century had dawned in the glorious certainty of Pierre-Simon de Laplace and the comprehensive universality of Newtonian mechanics. Its end, by contrast, was marked by a growing crisis of confidence in science and technology, a crisis matched in politics, philosophy and art – indeed in more or less everything.

The Industrial Revolution wasn't a revolution any more: it was a way of life – sometimes, as for many in today's globalized economy, a decidedly unpalatable one. Science had done great things, yes, but it had done questionable things too. Centuries-old communities displaced, giant filthy cities mushrooming haphazardly from the earth, and nightmares of suffering, slums and disease. Land-tied cultures that had lasted millennia dissolved away in a few decades of irresistible social change. Ever-growing populations, mercilessly exploiting empires, advancing materialism and dwindling faith in the old gods. In Europe the Great Powers of Britain, France, Germany and Russia continued on political, economic and imperial collision course, powered by thundering machines – and, of course, ever more efficient weapons of destruction.

It was no surprise if some commentators saw disaster ahead, and were more than ready to put the blame on science.

A philosophical difficulty

Fin de siècle. The fin, in fact, of a siècle that had promised unimaginable progress for all humanity. Much had turned out to be far more difficult than anticipated. Science and the material view, argued many, were simply proving inadequate to cure the world's ills. Humanity had strayed too far along the materialistic path.

---

was, like entropy, fundamentally concerned not with a static picture of reality, but with *change*.

What did the scientists themselves think? In a speech to com-
memorate the opening of the new Cavendish Laboratory in Cambridge in 1870, even the great James Clerk Maxwell counselled
against the hubris of the early 1800s. Maxwell pointed out that
whatever scientists' current confidence in their model of the Universe, it was unlikely that they had yet uttered the last word about
the nature of reality. Along with his insatiable curiosity, Maxwell
always showed a healthy respect for the complexity of nature. He
knew very well there was still a long way to go, and a lot that
could still turn out to be wrong.

He did still believe, even so, that science was on the right track:
it was using the right methods in the right frame of mind. Unfortunately, by the late 19th century there were many influential
thinkers who did not agree with such an optimistic view.

Science had reached a dead end, they said. The scientific
method had nowhere else to go, not because it had fulfilled its
promise to solve all the problems of mankind, rather because it
had reached its limits. And reached them far short of expecta-
tions. Science had let people down. The world was still full of
problems – fuller than ever, some felt. And science was stalled.

Naturally, certain philosophers began to suggest that there
were ways other than rational scientific enquiry to probe the
nature of reality. The Frenchman Henri Bergson, for instance,
proposed that a sort of *intuition*, instead of intelligence, could be
used to grasp more about the world. Bergson denied science's
technique of understanding whole ranges of phenomena through
single general theories. For Bergson, one had to comprehend phenomena one by one, to learn by an ever-increasing precision of
particular observation, an ever-deeper penetration into the singular nature of a thing. Unlike some 'anti-science' attitudes, this
wasn't a call to ignorance, a denial that knowing about the world
was an important human activity. Rather it was a claim that there
must be another way to know, to find out and understand, in
those regions of experience where experimental research was
failing.

The problem for many philosophers was that much of science lacked a proper notion of *time*. Nothing *developed*, things just stepped back and forth. Mechanical theories of forces saw the world as a set of static pictures, simply transforming one into another, like hopping between stepping stones. There was no sense in which that sequence of pictures had any developing 'meaning'. Nothing evolved, time had no fundamental role: things simply switched back and forth. The scientific Universe seemed to be a Universe without Time, without Becoming.

Philosophy's *fin de siècle* concern with time perhaps finds its greatest expression in the vast novel by one of Henri Bergson's staunchest followers, celebrated cork-wallpaper enthusiast Marcel Proust[2]. Proust's 3,000-page *oeuvre*, *A la Recherche du Temps Perdu* (literally 'In Search of Lost Time'), is, as you might expect, all about time. Its unnamed narrator delves into his own past, attempting a Bergsonian 'ever-deepening penetration' into the significance of his every experience[3]. Proust tries to get at the reality of a life that seems inextricably chewed up by Time. A reality that, according to the Bergsonian philosophy at least, science could offer no route to.

So despite the scientific successes of the previous century, science was in trouble. Like it or not, scientists have always had to work in a political environment. Then or now, in a society turning away from science scientists soon find it difficult to communicate their ideas, to fund their research, or to attract and teach new generations to follow them. Under attack from Bergson and others, science looked vulnerable.

There was more to it than philosophy. Science's real problem lay inside science itself: the still-invisible atom. It was like an itch

---

2 In an attempt to escape the distractions of the external world, Proust had his Paris apartment walled in soundproofing cork.

3 One publisher, in refusing to publish Proust's first volume, expressed mystification that a writer could spend the first 30 pages of a novel describing a character doing nothing but rolling over in bed trying to get to sleep.

that scientists still seemed unable to scratch. Was the atom real, or just a mathematically convenient fiction?

Fortunately, in the closing years of the 19th century a few scientists were going back to a puzzle that had lain unsolved since 1827. It was time for the middle world and Brownian motion to pop up again, and offer that elusive route to the atom.

Going back to Brown

Rather than explain Brownian motion, many scientists through the middle of the 19th century were keen simply to prove that it was all a big red herring. Reports claimed that it was just an effect of vibrations or temperature fluctuations. Against the backdrop of Boltzmann's statistics and the kinetic theory of gases, finally some turned again to Robert Brown's random dancing pollen, and began looking for a proper explanation.

One scientist who appreciated early on the significance of Brownian motion was French physicist and chemist Louis Georges Gouy[4]. Gouy was born in the Ardèche region of France in 1854, and worked in Paris and in Lyon. Chemists remember him for his trademark of exceptionally careful experiments, on matters such as how small particles in solution exert forces on each other and how liquids and gases become indistinguishable at the so-called 'critical point'. Happily, Gouy also became interested in Brownian motion, perceiving that it might hold a very special message about the nature of matter.

First Gouy had to prove to everyone's satisfaction – including his own – that Brownian motion was indeed a real phenomenon. He started by essentially going back half a century and repeating Robert Brown's experiments. For instance, to make certain that Brownian motion was not due to some external effect such as vibrations, Gouy even ventured out into the countryside, away from the rumbling cart and horse-traffic of the city.

---

4 Pronounced something like 'Gwie'; he is often known as 'Leon' rather than 'Louis Georges'.

In 1888 Gouy published the results of his experiments and put Brownian motion back on the menu of confirmed phenomena in the world of liquids. It seems disappointing that experimentalists had not really advanced at all in the 50 years since Robert Brown's observations of dancing pollen in the summer of 1827. But the road that science follows is rarely a straight one. Anyway, meaningful debate about the significance of Brownian motion was impossible before Maxwell, Boltzmann, Carnot, Thomson and Clausius.

On the basis of his experiments Gouy realized, at last, what the real contradiction was behind the random, incessant dance of Robert Brown's pollen. The contradiction, as Gouy put it in an 1895 paper in the *Revue des Sciences*, was this:

Brownian motion appeared to violate the Second Law of thermodynamics. It broke the unbreakable.

Perpetual motion – Brownian style

Gouy's argument went like this. If the random motion of the pollen particles is due to the atoms of the liquid buffeting the pollen around, this amounted to heat energy being used to move pollen particles around. The movement of objects by energy constitutes work. In Brownian motion, therefore, heat was being converted into work.

But the Second Law of thermodynamics states that in any process, only a certain fraction of heat can be converted into work. Some energy must always be lost. This is the principle behind William Thomson's heat death of the Universe. Now if in Brownian motion some fraction of the available energy is always discarded every time a liquid atom pushes a pollen particle this way or that, then eventually all the energy must be used up or lost to waste.

In other words, according to the Second Law of thermodynamics the random dance of the pollen must eventually stop.

That, as Brown and Gouy himself had shown, does not happen. Robert Brown had checked samples that were centuries old. They still danced.

Moreover, as Gouy put it, Brownian motion seemed to imply that that old miracle, the old dream of the alchemists, the *perpetuum mobile* or perpetual motion machine, was possible after all. Consider a tiny string connected at one end to the drive shaft of a motor, and at the other end to a tiny vane. Plunge the vane into a suspension of Brownian particles. As the perpetually dancing particles collide with the vane they rotate it, twist the string, and thus drive the motor. And because the Brownian dance never stops, they drive it *perpetually*.

Leon Gouy had no answer to this apparent conundrum. Maybe, he suggested, the Second Law of thermodynamics simply was not obeyed at the small scale of Brownian motion. After all, that law was based on experiments with large volumes of gas – or indeed on the theoretical statistics of very large numbers of 'atoms'. What if, on a microscopic middle scale, matter and energy did not obey the same rules?

In search of a theorist

Leon Gouy's perceptive comments indicated that Brownian motion represented experimental scientists' route into the heart of the contradiction between matter and energy, between mechanics and thermodynamics. Robert Brown had looked long and hard at his pollen particles, mineral grains, London soot and Sphinx dust. Gouy had gone back over the same ground to convince himself that Brownian motion was fundamental to middle world matter. The question facing experimentalists now was, what did you measure to learn more?

Post-Gouy, the demand was, in short, for maths that would predict numbers and experiments that would measure those numbers: a testable theory of Brown's dancing pollen.

Clearly, if you need a theory, then you need a theorist.

And anyone who needed a theorist, around the year 1900, could have done a lot worse than take a trip to Berne, Switzerland. There, if you looked hard enough, you might track down a certain young patent clerk. This rather obscure young man, despite the belief of most of his schoolteachers that he would never amount to anything, had been dreaming some big dreams. He had spent the last few years thinking seriously about the nature of reality. He was a young man who was about to change the world – not once but three times.

Albert Einstein was about to discover for himself the secrets of the middle world.

Chapter 7
A THEORY OF THE BROWNIAN WORLD

Over the course of the single year 1905, three pieces of research were published that changed our view of the Universe forever. They were all done by the same man, that obscure young patent clerk, Albert Einstein.

Revolutionary theory #1: Special relativity. Light is the only thing you can truly rely on in the Universe, the only thing that travels at constant, absolute, speed – with the consequence that in order to accommodate such rigid light, space and time must themselves become malleable, bendable, miscible, negotiable.

Revolutionary theory #2: The photon of light. Light energy, whenever it interacts with matter, behaves as if it came in small indivisible packets, or particles called photons. Following on from German physicist Max Planck's introduction of fixed or 'quantized' energies as a mathematical convenience, with this paper Einstein kick-started the quantum theory of subatomic matter.

And Revolutionary theory #3: Brownian motion.

Einstein's three theoretical revolutions of 1905 have fared rather differently in the popular consciousness. Almost everyone has heard of Revolutionary theory #1 – special relativity. Quite a lot of people know something about Revolutionary theory #2 – quantum physics and the photon. Few people are even aware of Revolutionary theory #3.

Which is, scientifically speaking, rather unfair. Without getting into a fruitless 'my favourite Einstein is better than yours' argument, suffice it to say that Einstein's Brownian motion work turns out to have at least as much relevance today as his relativity and quantum physics papers. In the century since Einstein's 'miraculous year', his papers dealing with the theory of Brownian motion have received by far the largest number of citations in other researchers' work.

So why is Einstein famous for relativity and not for Brownian motion? Perhaps because the significance of the revolution that Einstein's Brownian motion theory represented has taken a little longer to become clear. It is a revolution not on the scale of the whole Universe, nor on the scale of subatomic particles. It's a revolution in the middle.

Just as Robert Brown's observations in 1827 were the first careful, systematic experimental glimpse of the middle world, so Einstein's work was the first theory of the middle world.

Molecules made visible

Albert Einstein was born at a lucky time for a young man with an insatiable curiosity. The nature of the world he grew up in was ripe for complete reappraisal. In the years leading up to 1905 Einstein set himself the spare-time task of puzzling out what was wrong with reality.

As he saw it there were three fundamental things in need of working scientific theories: light, gravity and matter. Light and gravity defined the way matter interacted: they were forces[1], they controlled what stuff could do. With special relativity and the idea of photons, Einstein found new ways to conceive of light. Gravity – well, that would have to wait[2]. That left matter. With

---

1 Light, as explained by Maxwell's electromagnetic theory, is part and parcel of the electromagnetic force.
2 During the decade after 1905, Einstein created the formidable General Theory of Relativity: a more complete theory of light, space and time, including how gravity knitted the three together.

his Brownian motion theory, Einstein showed the way to proving the nature of matter.

Einstein's analysis of Brownian motion had statistics right at its heart. Einstein actually reinvented much of the statistical work of Boltzmann. That was Einstein for you: why use someone else's ideas when you can have your own? Statistics was the basic link between all Einstein's thinking leading up to 1905 – the theory of the photon also relied on statistics to interpret the interaction of light and matter.

Einstein's Brownian motion insight was this: what would really prove the statistical-atomic theory of matter was a way to *observe* the statistics in the flesh. Calculations weren't enough, even when they gave the right answers. That wouldn't satisfy those pesky philosophers, for who knew how many different sorts of calculation would give those same answers? You needed to *see* the numbers made real.

So Einstein found a way to lift the statistics of fast-moving invisible atoms and molecules up into the visible world under the microscope – into the middle world.

We don't know whether Einstein had read Leon Gouy's papers before 1905 – or even those of Robert Brown[3]. He went beyond Gouy's ideas by adding concrete maths, equations enabling numerical predictions that could be tested in experiments.

How, Einstein asked, can we describe the motion of a molecule in a liquid? The liquid is so crowded that no molecule can simply move in a straight line: it will continually collide with its neighbours and cannon off in some other direction. So let's assume, Einstein went on, that the path of the molecule is like a *randomly*

---

3 Some claim that Einstein didn't even know there was such a phenomenon as 'Brownian motion' when he published his first paper: his predictions had simply come straight out of the maths. It's notoriously hard to trace the footsteps of Einstein – the sources of his ideas. He was sparing when it came to citing others in his own papers. Letters indicate that he knew something about Brownian motion, though had perhaps not read much of the literature, before 1905.

varying quantity: in a long enough time interval the molecule suffers so many collisions that, from the beginning of one time interval to the beginning of the next, the molecule's speed and direction are 'randomized'.

Given the molecule's speed and direction at one time instant, because of the randomizing effect of so many millions of collisions, you might just as well roll dice to decide which way the molecule will be going at the next time instant. The molecule carries out a 'random walk', hopping now this way, now that – driven by the randomizing effect of many collisions.

Einstein constructed a mathematical description of this molecular random-walk motion[4]. Then he went further.

Now take a *large* particle, proposed Einstein, one much bigger than the molecules of the liquid. Place it in the liquid – immerse it in the sea of randomly jiggling liquid molecules. How does the large particle move?

The particle's motion, Einstein pointed out, depends on its energy. He knew from kinetic theory going back to Rudolf Clausius that the average energy of a body was given by its temperature. Know the temperature and you know the energy; know the energy and you can describe the motion.

So what is the particle's temperature? It must, Einstein realized, be the same as any of the *molecules* in the liquid: if not, the particle would either gain or lose energy – heat, essentially – until the temperatures were equal.

And if the large particle has the same temperature as the liquid, its average energy must be just the same as that of any molecule.

But that implied the large particle must behave just like a molecule. It would move like a molecule, jumping about randomly, buffeted around on the sea of ever-colliding liquid molecules. A

---

4   In fact he had already done this maths, before 1905, for a single molecule: he wrote it up in his PhD thesis, published at the beginning of 1906.

large particle immersed in a sea of invisible molecules would behave exactly as if it were itself a molecule! A 'molecule' large enough to be seen in the microscope.

It was molecular reality – made visible.

Einstein's theory showed that, *if* a liquid was composed of a sea of invisible molecules, a large visible particle immersed in it must execute a random walk – Brownian motion. Turn the argument around and you have experimental proof of the atomic nature of matter. Observations of Brownian motion implied that liquid matter *was* indeed made from billions of tiny particles. Boltzmann's detractors, who said direct evidence for his invisible atoms would never be found, were wrong. Brownian motion, the middle world dance of the pollen, was that very evidence.

It was a classic Einsteinian *gedanken*, a 'thought experiment' that cut to the fundamentals of the question – how can we *find out* what matter is made of? Einstein has always been revered as a theoretical physicist rather than an experimentalist. Though he may not have done many experiments himself, he did his thinking in the world of the experiment where things were observed and measured.

However, perhaps Einstein should not have exclusive credit. He wasn't the only one to find a theory of Brownian motion.

In March 2005, in a conference centre in a quiet corner of the Dutch city of Leiden, a small group of scientists gathered to celebrate 100 years of Einstein's theory of Brownian motion – and, more importantly, to discuss the latest consequences of those accidental observations of Robert Brown's way back in the summer of 1827. A German physicist named Peter Hanggi – well known for his work on Brownian motion in biological systems – stood up to speak. 'This conference is called "Brownian motion after Einstein"', he began in a determined voice. 'But perhaps it should be called: Brownian motion after *Sutherland*.'

The lecture room contained some of the world's leading authorities on all things Brownian. Next to no one had heard of Dr William Sutherland.

In the long shadow of Einstein: Sutherland and Smoluchowski

In 1862, the Sutherland family of Glasgow sailed down the River Clyde and out of Scotland forever. They were on their way to Australia. The youngest, William, was only three years old. The family settled in Melbourne. They prospered; the children, George, Alexander, Jane and William, received a good education. In 1879 William graduated with a First class degree in Natural Sciences from the University of Melbourne. Winning a scholarship, he travelled back to the other side of the world and did another degree at UCL in London, returning to Melbourne in 1882. He became a scientist of international repute; over three decades he published some 78 papers in the major journals. He acted as science correspondent for the local *Melbourne Age* newspaper (his brother George became its editor later on, in 1902). Having taught courses at the University unofficially for years, in 1888 William applied for the newly vacant chair of Natural Philosophy; due to an administrative blunder, his application was misfiled and he was never considered for the job; but despite that he even filled in as a stop-gap while the University waited for the eventually successful candidate, a physicist called T. R. Lyle, to arrive from England. Sutherland never did get a permanent, official position. He lived modestly with his sister Jane in a house in the Melbourne suburb of Kew, surviving off the family savings supplemented by whatever teaching work came his way.

In 1904 Sutherland published a paper in the London *Philosophical Magazine* that matches much of Einstein's analysis of the dynamics of motion in a liquid. Sutherland arrived at the same equation as Einstein for the rate at which a molecule diffuses in a solution. All this a year before Einstein's 1905 publication in the German journal *Annalen der Physik*.

Why has Einstein taken all the credit? For one thing, there is the obvious drama of Einstein's *annus mirabilis*: three great revolutions all in the one year. For another, the first decades of 20th century physics were dominated by the great universities and great scientists of the German-speaking world: Einstein himself, and Max Planck, and, later, the leading lights of quantum theory such as Erwin Schrödinger and Werner Heisenberg. Sutherland was known and respected: he was one of only two non-Germans invited to a 1906 conference celebrating the work of Ludwig Boltzmann, for instance. Perhaps Australia was just too far away from the centre of the scientific world.

Sutherland certainly deserves some recognition. But the game of 'who did what when?' in science is a notoriously difficult one – we've already seen how it distracted Rudolf Clausius at a vital moment in the development of 'his' kinetic theory. In the end, perhaps it misses the point. Despite the romantic image, scientists rarely work in intellectual isolation, nursing their revolutionary theories like alchemists jealously guarding the secret of the philosopher's stone. Many other researchers got very close to special relativity, for instance. Not even Einstein worked in a vacuum.

Sutherland's case demonstrates that as the 20th century dawned Brownian motion's time had come. The tide of science was pushing inexorably toward finding experimentally measurable consequences of Maxwell's and Boltzmann's statistics and randomness. Einstein was not alone in realizing the importance of the middle world in turning convenient mathematical devices into real things.

Einstein's rivals for the claim of being the first to reach a theory of Brownian motion don't end there. There was another scientist who was not quite so hidden in Einstein's shadow as Sutherland. This was Polish physicist Marian von Smoluchowski. Smoluchowski's name lives on in important equations describing processes of fragmentation and agglomeration – still used today in fields from fundamental physics to population growth to the engineering of chemical products like washing powders.

In 1906 Smoluchowski published his own theory of Brownian motion – a theory that he had actually completed a year before reading Einstein's paper. Smoluchowski came at the problem from a more mechanical point of view than Einstein, but nevertheless found the same result[5]. More importantly, Smo-luchowski was far less coy than Einstein in clearly stating the experimental situation he was trying to understand. He thoroughly reviewed the known facts of Brownian motion, setting his work clearly into context. Einstein's own analysis, by contrast, seems almost sly. It is possible to read it and almost miss what it's actually getting at. As in art and literature, different scientists often bring quite different perspectives to the same issues.

Whatever William Sutherland's and Marian von Smoluchowski's claims to equal rights with Einstein in the Brownian motion theory, mathematically speaking someone else beat all three of them. A mathematician studying the Paris stock market.

The random walk of wealth

In 1900 Frenchman Louis Bachelier submitted a PhD thesis, entitled *Théorie de la Spéculation*, to the University of Paris. In it, Bachelier analyzed a problem seemingly far removed from the middle world of dancing pollen. Yet, again, statistics and randomness were the key.

Bachelier was interested in the behaviour of prices on the *Bourse*, the Paris stock exchange. Stock price fluctuations could determine the future of whole industries. So how to predict the vagaries of the market? Such a question has worried many minds before and since. Bachelier was probably one of the first to turn to the statistics of random numbers.

Bachelier proposed treating prices simply as *random variables*, numbers that sometimes went up, sometimes down. The market as a whole thus comprised a large set of changing random numbers. Despite not knowing how individual prices might behave, you could still predict the average or overall behaviour of the market by computing the statistics of that set of fluctuating numbers.

5　At least, the same apart from a numerical factor that Smoluchowski later realized he had got wrong.

The mathematical link is clear. Einstein described the random hopping about of a middle world particle, Bachelier described the random movement of a stock market price. In both cases, calculating the statistics of these random walks led to equations that predicted their average behaviour: in Brownian motion, the typical distance travelled in a given time period by a Brownian particle; in Bachelier's stock market, the typical overall change in a price over a given number of days or weeks. Despite their completely different real applications, the two phenomena were described by the same equation.

Bachelier's equation was not a recipe for instant wealth. Just as Einstein's maths didn't tell you definitely where a particular particle would go, so Bachelier's theory did not tell you which stocks would go up and which down. It said, instead, something about the probability of various possible outcomes. The stock market was still a casino game; Bachelier had just made the odds a bit more quantifiable.

Bachelier did all this in 1900 – five years before Einstein's paper. There is no evidence that Einstein was aware of Bachelier's work, which is unsurprising for a man who possibly hadn't even heard of Brownian motion, let alone read an obscure French PhD thesis. In fact, Bachelier's work suffered from a lack of appreciation even amongst his native mathematicians. As late as the 1920s he was turned down for a professorship in Dijon because of the misguided criticism of one of the French maths establishment of the time, Paul Levy (who later realized his mistake and apologized to Bachelier).

Was Bachelier himself aware of the connection between his stock market maths and the behaviour of matter? Probably. Around 1910 he gave a course at the Sorbonne in Paris in which he discussed analogies between his financial work and problems in physics. As we shall see, his Sorbonne colleague Jean Perrin was carrying out the definitive experiments on dancing middle world particles at almost the same time.

Since Bachelier the science of random walks in economics has received vast attention – for obvious reasons given the amount of money at stake. Mathematicians and theoretical physicists working for banks, finance companies and stockbrokers routinely calculate the statistics of fluctuating money. Today, this financial Brownian motion has, like physical Brownian motion, moved into some exceptionally deep waters, trying to understand the complex ways that prices and markets interact, the long-term evolution of wealth and economies, and so on. At the heart of it all is still that simple insight: statistics and randomness hold the key to complex shifting entities, whether pollen or stock prices.

Albert Einstein, William Sutherland, Marian von Smoluchowski, Louis Bachelier... and indeed all the other Brownian motion theory pioneers that future historians of science may dig up: between them they put Brownian motion on a sound mathematical, thermodynamic and mechanical footing. And so to the year 1908, when one more scientist was ready to play a significant role in the transformation wrought by the middle world on the science of matter. Frenchman Paul Langevin was about to bang the final nail into the coffin of determinism. It was officially time to welcome Randomness back into the old world of Rules.

The Picasso effect

Remember Hieronymus Bosch and his last howling cry of mediæval chaos, *The Garden of Earthly Delights*? There's an odd parallel in the case of Paul Langevin. There exists a sketch of Langevin by Pablo Picasso, powerhouse of the Modern Art revolution. In a sense Picasso did to art what Langevin did to science: he let chaos back in.

Paul Langevin was born in 1872, into a humble family where education was traditionally an unaffordable luxury. Langevin nevertheless managed to get himself some, and by the 1890s was working under Pierre Curie at the Ecole de Physique et de Chimie Industrielle in Paris. In 1897 he crossed the Channel to work on X-rays alongside a young Ernest Rutherford, the future pioneer of atomic structure, at the Cavendish Laboratory in Cambridge. In 1905 Langevin returned to Paris to take over Pierre Curie's professorship. By this time Langevin was following lines of thought very similar to the still unknown Albert Einstein. He soon reformulated Einstein's analysis of Brownian motion, and he independently arrived at Einstein's realization that matter and energy were fundamentally interrelated by the famous $E = mc^2$. Langevin became something of a champion of Einstein in France, where simmering anti-Semitism and anti-German feelings had worked against the acceptance of Einstein's ideas[6].

In 1908 Langevin published a short paper in the journal of the French Académie des Sciences. In it he refers to Leon Gouy's perceptive comments on the significance of Brownian motion, as well as to Einstein's and Marian von Smoluchowski's analyses. Langevin pointed out that there is a third way to reach the mathematical expression of Brownian motion – 'by a completely different method, a derivation infinitely more simple'. He proceeded to write down quite the simplest yet most fundamental equation in physics: the celebrated Force equals Mass times Acceleration: $F = ma$. Here, m is the mass of a particle, F is the total sum of forces on that particle, and a is the resultant acceleration – speeding up or slowing down – of the particle, caused by those forces.

Langevin's goal was to get at the way a Brownian particle moved – its a, essentially – by writing down an expression for the forces on it – the F. One force on a Brownian particle – a pollen particle or any other dancing middle world inhabitant – was well known: the friction from the surrounding liquid. Just as there is friction when you rub yourself dry with a towel, or friction between a car and the air it speeds through, so a middle world particle dancing through water feels a frictional force trying to slow it down. Since the middle world particle *doesn't* stop, there

---

6 In 1898 Langevin signed the famous petition of novelist Emile Zola in support of the Jewish soldier Captain Dreyfus, unjustly jailed for treason as a result of anti-Semitic intrigues inside the French government.

must be, therefore, another force, call it X, kicking the particle, keeping it going against friction.

Force X is of course the ceaseless buffeting by the molecules of the liquid. Langevin's master-stroke was in realizing that X, to properly describe the random buffeting of the middle world particle, must simply be written as a randomly varying quantity.

A short interlude of mathematics later, and Langevin, starting from this most basic of physics equations, $F = ma$, arrived at the very same prediction for the 'statistical' motion of the particle as Einstein found[7]. Compared to Einstein's rather convoluted arguments about thermodynamics and statistics, Langevin's idea was startlingly simple. What was more, by putting the random force X directly into $F = ma$, he combined randomness and the bedrock of Newtonian mechanics explicitly – to reach the equation of the middle world.

What became known as the 'Langevin equation' is indeed for most purposes 'the equation of the middle world'. Textbooks introducing today's middle world physicists, chemists and biologists to the theoretical analysis of Brownian motion invariably use Langevin's mathematics rather than Einstein's or Smoluchowski's. What many of these students don't realize is what Langevin's approach meant for science in 1908. It was the unashamed demolition of the wall built over generations between the Garden of Earthly Delights and the Orchard of Determinism. Langevin's merging of middle world randomness and $F = ma$ finally brought the wall crashing down.

The end of the affair

In 1906 Paul Langevin's teacher and mentor died in a traffic accident. This was Pierre Curie, Nobel Prize winner, leading light of French science of *la belle époque*, husband and collaborator of

---

7   In fact, Langevin's solution is slightly better than Einstein's, as it works over shorter time intervals, describing the random dance with greater precision.

Marie. Returning from a scientific meeting one evening – perhaps engrossed in scientific thought – Pierre stepped absently into the rue Dauphine, near the Pont Neuf, just south of the Seine. He was run down and killed by a horsedrawn cart.

He was only 47. It was a blow to French science and a tragedy for Marie and the Curies' two daughters, Irène aged nine and Eve aged only two. Marie and Pierre had been married 11 years. The devoted couple had been the centre of a thriving social community of physicists, chemists, biologists and other intellectuals – a 'circle' not too dissimilar to the more celebrated ones of the Parisian art world of the time. The circle included Paul Langevin and another soon-to-be pioneer of the middle world, Jean Perrin.

After her husband's death, Marie threw herself into the unconventional education of her daughters. The Curie girls, along with the children of other members of the circle, such as the Perrins, were shepherded back and forth between Perrin's Sorbonne labs for chemistry lessons, Langevin's home for maths, the studios of a sculptor called Magrou for lessons in art, and Marie's own classrooms to be taught physics. Marie, an opponent of traditional schooling, thought that children needed lots of running about in the fresh air as well as time spent in schoolrooms poring over textbooks. (Meanwhile Marie herself, labouring away in the lab, ignored the growing health warnings about radioactivity, as had her husband.)

Somewhere in all this social, educational and scientific whirl, widow Marie's relations with Paul Langevin shifted – from friend to lover.

Langevin was already married with four children. The marriage was well known to be failing; Langevin lived separately in a bachelor apartment. Nevertheless, in 1910 one did not embark on such an affair lightly. Marie Curie was famous. Famous as a Nobel Prize winner, famous as the bereaved wife of the great Pierre. Famous as a female scientist. She was an obvious target for the celebrity treatment even back then.

In 1911, news of the affair leaked out after Langevin's furious wife Jeanne engineered the theft of some letters and arranged their delivery to a journalist. It began to look as if Marie Curie's reputation would be utterly destroyed.

Scientists today sometimes complain that they seem to be invisible to the media, their work and lives ignored. Judging by what happened to Paul Langevin and Marie Curie, perhaps they should be glad. Langevin, as a distinguished professor, wasn't supposed to get involved in scandals like this. (It was perfectly acceptable for French gentlemen to have mistresses. Langevin's real crime in the eyes of society was that his mistress wasn't some invisible other woman – she was a celebrity in her own right.) Marie Curie, as a woman and, significantly, not a *French* woman (she was Polish, her maiden name being Sklodowska), certainly should not be dragging away the husband of a respectable French lady with her six[8] children. Marie was vilified, especially in the right-wing press, for corrupting French science, French men and French families. How could the good French stock fight against falling population (an eternal obsession of the French as they worried about the mushrooming population of Germany) if their husbands were stolen by foreigners?

At one point Marie contemplated suicide, and later had to be hospitalized with a serious kidney infection exacerbated by stress. Langevin meanwhile ended up fighting an absurd duel with an old schoolfriend turned rabid journalist. The journalist, claiming patriotically that he could not bear to deprive Jeanne Langevin and her four[9] children of their support by killing Langevin, refused to fire. There followed a hurried debate amongst the assembled seconds and gaggle of journalists, ending with both parties firing harmlessly into the air.

---

8 Journalists of the day, in the time-honoured tradition of neglecting reality when it suited, gleefully inflated the number of the Langevins' children by 50%.

9 By now the number had gone back down to four.

Even the award to Marie of her second Nobel Prize in November 1911, this one for chemistry, was not enough to deflect the criticism. Though friends such as the Perrins rallied round with continuing support, Marie and Paul's love affair was doomed. Marie threw herself into science, planning and creating a new 'Institute of Radium' that eventually contributed widely to medical uses of radiation. Marie trained doctors in the use of X-rays during the First World War. Langevin, plunging into war work, invented a method of using sound to detect underwater objects – sonar. It was even tried out as a weapon, Langevin demonstrating how it could be used to kill fish. Years later, Langevin was still living apart from his wife, though he was reported to have taken another mistress: this time a satisfactorily invisible one, an anonymous secretary. The morals of the Third Republic were OK with that.

There is a final link between Langevin and Curie. In 1943, as the occupying Nazi forces closed in on outspoken anti-Nazi scientists such as the 78-year old Langevin, Frederic Joliot-Curie, husband of the Curies' daughter Irène, arranged Langevin's escape out of France and safe passage to Switzerland. Irène and her children went with Paul. Marie, of course, was long dead: succumbing to leukaemia in 1934, after years of exposure to radiation.

Meanwhile, the experiments...

One might say that, despite the use of mathematical randomness in the Langevin equation, nothing had really changed. Underneath the fluctuating statistics of atoms, the maths of Boltzmann, you still had deterministic particles obeying Newtonian mechanics. So was randomness really necessary?

Well, ask a working physicist today if he thinks he could do physics without the Langevin equation, without the 'fluctuation-dissipation theorem' that is the basic product of Einstein's analysis; without the vast array of mathematical and conceptual tools that the treatment of fluctuations and randomness has provided.

Even the later development of quantum mechanics, the science of the subatomic world, benefited significantly from the middle world work of Einstein, Langevin and the rest. The science of matter today (not to mention the science of finance) would be impossible without randomness. Just as the art of today – whatever your opinion of it – would simply not exist without Picasso and the 'modern' movement that arose at the turn of the 20th century.

But we are getting ahead of ourselves – we are forgetting one of the cardinal rules of doing science. So Einstein and Smoluchowski and Sutherland and Langevin have come up with theories for Brownian motion. Theories are never enough in science. *Nullius in verba*, as the motto of the Royal Society goes: which loosely translated means 'take no-one's word for it until you've seen it for yourself!'. The first members of the Royal Society, one of the crucibles of modern Western science, prided themselves on the concept of experimental proof. For too long thinkers had expounded clever ideas, twisted the minds of whole populations, and turned out to be deluded dreamers. What a proper understanding of reality demanded was solid experimental evidence.

And this was no less true just because one of the theorists in the case of Brownian motion happened to be the great Albert Einstein. What was still lacking, as Langevin published his short paper of 1908, was experimental proof: direct measurements from the middle world.

It was time to get out the microscope again, and go back to the methods of that long-dead, near-forgotten botanist, Robert Brown.

Chapter 8

FROM RUBBER BALLS TO ATOMS

Scientists are often amongst the first beneficiaries of new technology. Long before lasers were employed in cheap CD players, physicists were using them to study the intimate details of the structure of matter, and even to manipulate objects. Long before there was a computer on everybody's desk, chemists were using them to calculate the behaviour of atoms in a crystal, and mathematicians to crack enemy codes. And long before Hollywood turned films into a billion-dollar business, scientists were using movie cameras to study the middle world.

One of the first to catch the middle world on celluloid, in the opening years of the 20th century, was French physicist Victor Henri. Henri read Einstein's and Langevin's papers on Brownian motion and decided to have a go at validating Einstein's theoretical results.

Henri assembled a state-of-the-art set of microscopy kit, including the latest compound lenses, and combined it with no less advanced cinema equipment. Immersing some tiny spheres of rubber only a thousandth of a millimetre across in droplets of water, he began to capture Brownian motion on film. By measuring the particles' positions from frame to frame on the film, Henri could make direct measurements of their motion.

Different particles travelled different distances – as expected from a statistical model of matter. Henri calculated an average distance over a dozen measured particles, and using values for the viscosity of the water and the temperature, plugged his measurements into Albert Einstein's formula.

And the formula was wrong. His particles were going about four times too fast.

If science were a nice tidy story, and not the litany of surprises, unsolved mysteries, cack-handed mistakes and startling genius that it is, I could now reveal the cause of Henri's disagreement with Einstein. Unfortunately no one seems to have come up with a completely convincing explanation of Henri's results – except that, as Henri himself declared in the paper he published in 1908 in the *Comptes Rendus de l'Académie des Sciences*, measuring Brownian motion was a stiff challenge. The best candidate for experimental error in this case is that the bright light required to make Henri's primitive movies, concentrated into the focus of the microscope, heated up the water droplets so much that the measurements ceased to be reliable. This was Einstein's opinion, when he heard of Henri's measurements.

Had the matter rested with the experiments of Victor Henri, perhaps history would have been different. That reckoned without another French scientist with a trusty microscope of his own, and a burning desire to prove once and for all the nature of matter. It was the turn of Jean-Baptiste Perrin.

The man who proved atoms

Perrin, a physical chemist at the Sorbonne in Paris, had grown up, like Einstein and Langevin, on a diet of atomic theory. He remained completely determined to prove what he already believed: that matter was made of atoms.

Jean Perrin was a member of that close-knit circle of scientists, intellectuals and artists that buzzed around Paris in the last decade of the 19th and first decade of the 20th centuries. Born in 1870 in Lille, he had barely known his father, an army officer who was killed during the Franco-Prussian war of that year – the same war that Rudolf Clausius was engaged in, on the other side. Jean and his two sisters were brought up by their mother, and Jean eventually won a place at the prestigious Ecole Normale Supérieure in Paris. In the 1890s he studied in the same *milieu* as

Pierre Curie and Paul Langevin at the Ecole de Physique et de Chimie Industrielles. As we have seen, he was one of the friends who, in 1911, rallied round Marie Curie and Langevin as they came under attack from the newshounds.

Jean Perrin became a significant force in French science. He was largely responsible for the creation in the 1930s of the Centre National de la Recherche Scientifique, the body that still today runs publicly funded science in France. Perrin was also keen to push science out to the wider public, writing popular books, giving public lectures and setting up the Palais de la Découverte, Paris's science museum and hands-on education centre.

Like the Curies, Perrin and his wife Henriette brought up children who later followed their parents into science. His son Francis became a successful physicist, playing a key role in the science of atomic power and, of course, bombs. Jean Perrin had a passion for the social relevance of science, believing that the route to social development lay with scientific understanding. He believed too that science had to be a human activity, not some automatic machine that subjected whole populations to often unwanted social changes. Science, in other words, with a conscience.

Perrin became a committed socialist, and in 1936 convinced the new Prime Minister Léon Blum that the French cabinet must have a scientist on board. Science was a matter that governments could no longer ignore. Perrin himself served time on Blum's cabinet. As an outspoken anti-fascist Perrin was forced to flee France in 1940 for exile in New York, where his son Francis was already working as a lecturer at Columbia[1].

Invisible rays

But back to the 1890s. Before venturing into the middle world, Jean Perrin had a fair career under his belt studying an apparently unrelated subject.

---

1   Jean Perrin never returned to France: he died in New York in 1943, his country still under Nazi occupation.

Near the end of the 19th century mysterious invisible rays became the hot topic in *fin de siècle* physics. Kicked off by the accidental discovery of X-rays, suddenly a whole universe of previously invisible, completely unsuspected strangeness opened up, pulling the carpet out from under the feet of some rather too complacent physicists.

Despite the debates over statistical mechanics, kinetic theory, Newtonian determinism, the existence of atoms and the like, physics near the end of the 19th century had seemed to many to be almost finished. According to J. J. Thomson, head of the Cavendish Laboratory at Cambridge in the 1890s, many young trainee physicists of his generation wondered whether it was worth setting out on a career in physics. Would there be any jobs for them once physics was all wrapped up? Similarly, in Germany in the 1880s, Max Planck, future Nobel laureate for his work on quantum phenomena, was famously advised not to study physics at all. According to his teacher the subject was all but done and dusted.

The discovery of X-rays exploded that complacency. X-rays were just one in a growing family of strange invisible rays that invaded the pages of the science journals. Jean Perrin's interest had already been absorbed by another: cathode rays. These were emitted by a negatively charged 'cathode', detectable because they gave out bright light when they hit a screen. Cathode rays had first been discovered in the 1870s, but were still a puzzle 20 years later. In 1895 Jean Perrin's experiments revealed an important clue: cathode rays could be guided by applying electric voltages. Because of this, cathode rays became the secret of the first television screens.

Perrin also realized that by measuring the energy given to the rays as they passed through a known electric voltage, one could obtain a value for the rays' negative charge. By now it was clear that the rays were actually streams of particles. Measuring their charge would pin down what these cathode rays really were. Quickly Perrin set about constructing an experiment.

Not quickly enough as it turned out. On the other side of the Channel, J. J. Thomson at the Cavendish was also experimenting on cathode rays. Thomson had the same idea as Perrin, and designed an experiment by which, accelerating the rays via an electric voltage, he could measure the ratio of the charge on the particle to its mass. He showed that the cathode ray particle had a mass almost 2,000 times smaller than the hydrogen atom. It was an entirely new particle, a tiny component of the atom, which Thomson christened the 'electron'. This was the discovery that won Thomson the Nobel Prize. Perrin was pipped at the post by Thomson.

J. J. Thomson and his up-and-coming student, the New Zealander Ernest Rutherford, were amongst those counted as visiting members of Perrin's 'Paris circle'. The circle consisted for the most part of already convinced 'atomists'. In 1901 Perrin proposed a cartoon model of the atom: a central nucleus orbited by planetary electrons, a model that Ernest Rutherford and Danish physicist Niels Bohr later reinvented and turned into a real theory of the internals of the atom.

But when it came to the middle world, Perrin's close association with Paul Langevin was probably the most important factor. As part of his chemistry course at the Sorbonne, Perrin had been working on suspensions of middle world particles since 1903; it was through Langevin that he found out about Einstein's analysis of middle world motion, and was put onto the possibility of a direct experimental comparison.

Begun around 1907, Perrin's conclusive series of middle world experiments continued for more than five years and involved staggering efforts of observation and data collection. Despite the numerous remaining would-be doubters of atomism, Perrin's work proved impregnable. He eventually came to be called 'the man who proved atoms', and eventually collected his own Nobel Prize.

The perfect rubber ball

Still, in 1907 it was not yet time to start dreaming of prizes. Before he could do any serious experiments, Perrin had to find the ideal inhabitant of the middle world: he needed the simplest, most perfect, Brownian particle.

Question: if Jean Perrin wanted to study Brownian motion why didn't he simply follow in Robert Brown's footsteps, and go scrape up some pollen grains?

To prove anything, Perrin was going to have to measure some real numbers and compare them directly with Albert Einstein's equations. Perrin's particles would have to satisfy the stringent conditions of the theory. That was, to use a modern phrase, a big ask, and pollen particles weren't up to the job.

First, the particles would have to behave themselves. They must not attract nor repel other particles, nor stick haphazardly to the walls of the microscope slide, as such effects could significantly change the way they moved. Because the particles would be immersed in water, which contains electric charges, avoiding long-distance forces such as electrical repulsions between particles was not easy.

Next, the particles would have to be as spherical as possible. Pollen particles are more often tiny tubes or cylinders. Einstein's theory demanded perfect spheres – only that way were the calculations straightforward.

Thirdly, there was the density problem. The particles needed a density close enough to that of water that they would neither sink straight to the bottom nor float straight to the surface before Perrin had time to measure their Brownian motion.

After these three difficult enough demands, now came the really hard bit: size. The bigger the particle, the more it resisted being pushed around by the surrounding atoms of the liquid, and the slower it moved: the dance became more of a waltz than a salsa. Particles too big would have such slow Brownian motion that it would take months-long experiments to obtain data.

On the other hand, particles too small would move so fast that it would be impossible to track them reliably. If he was going to succeed, Perrin needed particles that fell squarely into a precise and rather narrow size window.

And lastly, the *really* really difficult bit. He needed a sample of particles in which all the particles were as near as possible the *same* size. If the rate at which particles danced at depended on their size, as Einstein's theory predicted, then a sample of particles of all different sizes would dance at all different speeds. There would be no easy way to verify the precision of Einstein's relation between dancing speed and size. The data would be all over the place.

In summary, measuring Brownian motion demanded the following: uniform, well-behaved particles that were just the right shape, size, density and stickiness.

Despite the stringent conditions of the theory, Perrin did discover a substance that more or less fitted the bill. It was a resin called *gamboge*, derived from a species of rubber tree. When the resin was mixed with water it formed little balls of just the right density that were stable enough to carry out long-term experiments, and whose interactions were simple enough that the theoretical conditions would be just about satisfied. By adjusting the conditions under which the particles were made, Perrin could also obtain particles just about within the narrow window of size required for the best data.

But control of size remained the biggest stumbling block. As careful as Perrin was, the *gamboge* spheres invariably still came out with quite a wide range of different diameters, spread around the required size. Perrin's challenge was to somehow remove the particles that were too big or too small.

Perrin needed particles around a thousandth of a millimetre across. Even a few cubic centimetres of suspension of particles of this size will contain several thousand billion particles. So selecting the best particles one by one is a non-starter. Even the most exploited PhD student in scientific history has never been landed with quite such a task.

Instead, Perrin used brute force. He put his suspension of 'raw' *gamboge* particles – all different sizes jumbled up – into a centrifuge. By spinning around in a circle very fast, the centrifuge generates large forces that push its contents outwards from the centre of the circle. This is the principle of the 'Wall of Death', the old fairground show where people ride motorbikes around the inside of a huge vertical cylinder – and manage to stay up even though it appears that gravity ought to make them fall down.

Inside the centrifuge Perrin's particles were forced outwards as it spun around. Crucially, the rate at which the particles moved outwards depended on their size. The bigger ones moved outward faster than the smaller. So after a session in the centrifuge the bottom of the sample tube contained mostly the biggest particles, the top mostly the smallest. By multiple repeated spins, Perrin separated his particles into different sizes.

It was a painstaking business. From a half-kilo of suspension he got just a few grammes of useable particles. But at last he was ready to get the measure of the middle world.

The middle world atmosphere

Perrin began by trying the same sort of experiments as Victor Henri: tracking dancing particles one by one, trying to measure how far they danced. But the experiments were too difficult to get any reliable data from – just as Henri had found. Perrin changed tack.

He started to think about gravity instead. Part of Albert Einstein's original analysis, and indeed a situation that went back as far as Ludwig Boltzmann's earliest statistical work on kinetic theory, was what a suspension of particles would look like if subjected to the downward pull of gravity.

Boltzmann was of course thinking about the atoms of kinetic theory, not middle world particles. Boltzmann's statistics predicted that in a gas subject to gravity, the population of atoms gets less as you go higher and higher. Fewer atoms have enough energy to resist the force of gravity and stay up near the top.

This is exactly what happens in the Earth's atmosphere. Climb Mount Everest and the air gets thinner and thinner.

Einstein's theory implied that this progressive thinning out of a gas of atoms with increasing height must also apply to large particles suspended in a liquid. Einstein's large particle was supposed to behave just like a giant atom after all.

Perrin's suspensions of *gamboge* particles were of course subject to gravity, just like anything else on Earth. So, Perrin realized, he could simply count how many particles there were at different heights in his suspension under the microscope: if Einstein was right he should find a tiny 'atmosphere' of middle world particles, thinning out as he counted higher above the bottom of the suspension.

Perrin's first paper, published in 1908 in the *Comptes Rendus de l'Académie des Sciences*, confirmed the prediction. Looking higher and higher above the bottom of the sample under the microscope, Perrin counted fewer and fewer particles[2]. Numerically, Perrin's measures matched the predictions of Einstein perfectly.

This was a much easier experiment than trying to measure distances travelled by particles, as Victor Henri had attempted. All it required was to count particles at each height. In practice Perrin still had to do many counts and many experiments to be sure: such is the nature of statistics. Nevertheless his results, in his own words, 'made the world of molecules more tangible' – at last.

Magic numbers from the middle world

Gaining experience with the microscope, Perrin and his Sorbonne students steadily advanced in the great task of data collection. They counted tens of thousands of particles to fix with as

---

2   Because middle world particles are much bigger, they are pulled down more strongly by gravity than an atmosphere of atoms or molecules like the air. The middle world 'atmosphere' is correspondingly squeezed into a smaller range of height. In terms of an equivalent drop in pressure, a climb of a few millimetres in Perrin's *gamboge* suspension is like climbing to the top of Mount Everest.

much precision as possible the statistics of the middle world atmosphere. Eventually they mastered tracking and measuring particle motions too. Instead of using movie film Perrin projected images from the microscope onto a wall, large enough for particle positions to be traced and recorded 'live' by the experimenter. Encouragingly, Perrin's results showed much better agreement with Einstein's theory than had Victor Henri's.

Perrin also did experiments with other substances, such as mastic, to check that the agreement with Einstein didn't vary with different suspensions. He developed three new ways to measure the particle size – an accurate knowledge of which was vital for precise comparison with theory. One of Perrin's students, René Constantin, measured how the particle population in a fixed region of the suspension fluctuated over time, as a result of the combined Brownian motion of all the particles. Constantin's results matched the predictions of Marian von Smoluchowski. The weight of data began to be almost irresistible, even for the most entrenched anti-atomist.

The final convincing proof came as Perrin began to fit his findings into a latticework of other scientists' arguments in support of the reality of atoms. This was a measure of something called 'Avogadro's number'.

A single thread ran through all Perrin's experiments, connecting them to other atomist theories of phenomena as varied as radioactivity and radiation: Avogadro's number. Named after Italian chemist Amedeo Avogadro, this universal figure counts the number of atoms or molecules in a fixed volume of gas. Avogadro suggested at the beginning of the 19th century that, if matter is made out of many tiny discrete particles, then a given volume of any simple gases should contain the same number of those particles, regardless of the chemical nature of the gas. He was at first largely ignored, but subsequent developments in chemistry and kinetic theory pushed 'Avogadro's number' to the forefront.

This number wasn't something you could just count. By all estimates it was stupendously large, around a hundred thousand billion billion. (Since atoms are small, any reasonable volume of gas will contain a *lot* of them.) Einstein and Perrin offered another way to measure this number, via the statistics of Brownian particles.

Perrin obtained five separate measurements of Avogadro's number, by different methods, in his middle world experiments. The measures were all in good agreement. In each case he got about six hundred thousand billion billion. He then compared his measurements to other completely different experiments. These ranged from the diffraction of light by gases, the scattering of light that makes the sky blue, the energy emitted by 'perfect radiators', the electrical charges on microscopic dust particles, and even from radioactivity, the domain of his old friends the Curies.

All these different phenomena gave comparable values for Avogadro's number, from six to seven hundred thousand billion billion.

The agreement, coming as it did from such a wide range of apparently independent fields of experiment – from Brownian motion to light to radioactivity – was the most compelling proof yet that the kinetic-statistical theory of Boltzmann and Einstein was sound.

In 1912 Perrin published a book summing up the arguments for the reality of atoms. *Les Atomes* sold 30,000 copies in its first year – and there was no going back. After more than 2,000 years of speculation and controversy, from Leuccipus, Democritus, Epicurus and Lucretius to Newton, Maxwell, Boltzmann and Einstein – atoms were real. And it was in the neglected middle world that Jean Perrin proved it.

At the time of Perrin's experiments it was a little over 80 years since Robert Brown first peered into his microscope at some *Clarkia pulchella* pollen, that summer's day in 1827. In 1926 – just shy of a century after Robert Brown's experiments – Jean Perrin was awarded the Nobel Prize for his work on Brownian motion.

So after years in obscurity, surely now the middle world had proved itself worthy of serious scientific limelight.

Unfortunately things didn't quite work out that way.

Scientists are never satisfied. While Perrin was busy proving once and for all that stuff really was made of atoms, others were already probing deeper, going right inside the atom. The Greeks may have declared their 'atom' as the 'uncuttable' bottom line, but already J. J. Thomson's cathode ray experiments had proved that it wasn't the smallest of the small. Atoms had sub-components, such as Thomson's electrons and yet more smaller constituents revealed by subsequent experiments.

In the process the sub-atomists uncovered something very odd. Very small particles such as electrons do not behave in any sense like the tiny billiard balls we might expect them to be. At this sub-atomic scale, the nature of reality changes. Science had stumbled on the quantum world. With big consequences: the mysteries of the subatomic created an almighty diversion that once again buried the middle world from all but a minority of scientific prying eyes. The diversion would last the best part of a century.

Middle world deserters

Subatomic entities like electrons do not behave as if they are single well-defined particles. Albert Einstein was one of the people who kick-started the science of this weird quantum world. Einstein's Revolutionary theory #2 suggested that a beam of light must sometimes be thought of as a set of particles or 'photons' rather than a wave. Similarly, a 'particle' such as an electron should sometimes be thought of as a wave, spread out across a region of space rather than with a single measurable position.

Anything small enough – an electron, a photon, or one of the many other species of subatomic particle discovered through the first decades of the 20th century – can only be understood properly by taking account of this peculiar quantum reality.

A mathematical theory of quantum phenomena, so-called quantum mechanics, was developed in the 1920s, championed by the Austrian Erwin Schrödinger and the German Werner Heisenberg. Quantum mechanics proposes that one must think of a quantum particle like an electron not as a single definable particle at all, but as a *statistical* entity.

As with any statistics, you can only calculate probabilities, not certainties. With quantum mechanics you can only calculate the probability of the electron being in this or that place, moving with this or that speed. You have to actually do the measurement to get a definite answer.

The middle world is a world much bigger than atoms, far above the scale where quantum effects are noticeable. Hence Brownian motion can be understood without recourse to quantum mechanics. Nevertheless, understanding the middle world was vitally important for the historical development of quantum mechanics. After Jean Perrin turned Einstein's statistical theory of Brownian motion into hard reality, the scientific world was far more ready to accept a statistical interpretation of the electron and other quantum particles[3]. It was official: statistics were good.

The rise of quantum science underlines a painful irony in the story of the middle world. The second and third decades of the 20th century were a time of furious excitement for physicists. The puzzles of the quantum world were so enticing that the vast majority of them deserted the middle world. They stepped down into the subatomic – dashing off in search of quantum secrets.

New subatomic mysteries were uncovered almost every day, and new theories and explanations followed no less rapidly. The Frenchman Louis de Broglie, the Austrian Erwin Schrödinger, the Germans Werner Heisenberg and Wolfgang Pauli, the Englishman Paul Dirac and the Italian Enrico Fermi – an apparently

---

3 Though the statistical maths of quantum behaviour works perfectly, even today there remains no satisfying conceptual interpretation of quantum mechanics. Einstein himself never felt comfortable with the idea. Not because he didn't like statistics (a common misconception of his attitude), but because quantum mechanics cannot say *why* a single electron behaves like a statistic. The purpose of statistics is to count things. So what, in the case of an electron or photon, are the statistics counting?

endless supply of startlingly clever young physicists plunged head-first into the quantum world – and each of them came up clutching unimagined treasures[4].

Meanwhile, Einstein himself went in the other direction. He stepped *up* from the middle world. His great quest – to understand light, matter and gravity – still lacked that last member of the trio: gravity. Between 1906 and 1916 he developed what many see as his greatest triumph, the General Theory of Relativity – the description of gravity as the shape of space and time. He spent the rest of his life searching for a way to turn three into one, to put light, matter and gravity together into a single unified theory. Einstein left the middle world so far behind that much later, writing his autobiographical notes in the 1940s, he dismissed his own work on Brownian motion as relatively unimportant.

Post-Perrin, with most physicists falling in love with the sub-atomic and Einstein never a crowd-follower, falling in love with the whole Universe, the middle world went back to near-obscurity. That strange place of restless dancing objects was quickly forgotten by those who had so recently used it to such great effect: to prove the reality of the atom, the nature of matter.

But it was now time for the middle world to provide equally profound clues about that other essential component of the Universe.

So much for matter. It was time for life.

---

4   In his doctoral thesis, Louis de Broglie proposed the 'matter–wave' idea that inspired Erwin Schrödinger's theory of quantum mechanics. De Broglie studied at the Sorbonne: his PhD examination committee included Jean Perrin and Paul Langevin.

Chapter 9
FROM MATTER TO LIFE

There's something fishy about DNA.

DNA, or deoxyribonucleic acid, is the molecule at the centre of life on Earth. DNA is the memory store of life. It encodes, as a string of chemicals known as a genome, the information necessary to reproduce an entire organism. We now know the complete genomes of a myriad of creatures, including plants, mice, monkeys and, of course, humans.

The first explanation of the structure of DNA was published in the journal *Nature* in 1953, by two scientists working in Cambridge, UK: an Englishman, Francis Crick, and an American, James Watson. They weren't working in a biology lab but in the Cavendish Laboratory, the centre of physics research at Cambridge. The structure or shape of the DNA molecule, as opposed to what it does biologically, is really physics.

Crick and Watson drew the clue that put them on the right track from data gathered by Maurice Wilkins and Rosalind Franklin, working at King's College in London. From Wilkins' and Franklin's indirect pictures of how X-rays were deflected by the various atoms in the chemical structure, Crick and Watson came up with a mathematically defensible 'guess' that the actual spatial arrangement of the DNA molecule is a twisted ladder. A double helix.

And there was much rejoicing. Crick, Watson and Wilkins got Nobel Prizes[1]; the rest of us got, basically, almost all of modern biology and genetics.

---

1 There was nothing for Franklin, for long before the Nobel she had died of cancer, at the tragically young age of 37.

There was, nonetheless, something rather artificial about the whole business. The DNA that Wilkins and Franklin measured and whose structure Crick and Watson deciphered wasn't alive and kicking. To make those X-ray pictures, the DNA had to be kept still. It was frozen life.

Frozen helix, wiggly chain

The structure of DNA is a double helix. But does that tell the whole story about the DNA molecule?

If you ask someone like Charlie Laughton, a scientist at the University of Nottingham, the answer is no, not really. Laughton and his colleagues use atomic-scale computer models of DNA to calculate how the molecule behaves physically, taking into account the fact that in the cell DNA is immersed in a sea of water molecules. Laughton's animated videos show not a prim helter-skelter sitting still waiting to be admired – but a twisting, writhing molecular beast. Most undeniably DNA alive and kicking. DNA that cannot stand still.

Why are Laughton's DNA models – and a lot of real bio-molecules – not quite the tidy static structures à la Crick and Watson? Because life is lived, it turns out, mostly in the middle world. The world where nothing can stand still.

As so often in this tale, size means everything. So how big is the DNA molecule? This is a trickier question than it sounds. Ask most people how big a typical 'molecule' is and they'll think of something only a bit bigger than an atom – say a molecule of water, two hydrogen atoms hooked onto one oxygen. That would be about a tenth of a nanometre across – one tenth of one billionth of a metre. Small, in other words. Not really middle world material.

But molecules don't have to be small. DNA is in one sense a giant. It is a chain of chemicals (called bases) that, if uncoiled, reaches up to a millimetre long.

Stored in the cell nucleus, DNA is wound up into a complicated ball, a fuzzy scrunched-up globule a few tens of nanometres across. It is surrounded by water. DNA is, in short, an inhabitant of the middle world.

No wonder it isn't really a frozen helix. Crick and Watson's structure of DNA has helped implant a false picture in our minds: DNA as a beautifully sculpted piece of chemistry. Crick and Watson's structure is not wrong: it's just incomplete. If DNA is a sculpture, it is a living, dynamic, eternally restless sculpture.

At the size of a few DNA units, a few bases, the structure *is* the famous double helix. At this scale chemical interactions are strong enough to resist the buffeting of water molecules – the inevitable random jiggle of Brownian motion. This is vital for the purpose of DNA, to safeguard the genetic memory. If the sequence of bases wasn't stable on a small scale, if it kept getting torn apart and rearranged, genetic information could not be reliably stored. Evolution wouldn't work and we wouldn't be here. So hooray for Crick and Watson's double helix.

But on a larger scale – sequences of dozens or hundreds of bases – the impact of local chemical interactions between individual bases becomes more and more feeble. At this scale, much bigger than the buffeting water molecules that surround the DNA chain, sections of the molecule inevitably wobble and pitch and flail around, in the grip of Brownian motion.

If DNA is the heart of life, then the heart of life is eternally restless.

The middle world comes alive

Life isn't only about DNA. There are many other processes and molecules fundamental to life where scientists have begun to realize that middle world behaviour holds the key.

The basis of life as far as we understand it today is the cell. Organs are partitioned into cells: water-filled compartments within membranous walls. Such separation into compartments is vital to establishing locally controlled environments, where processes necessary to the overall functioning of the living being can occur in some orderly fashion. The cell is a sort of city-state, shielded by the membrane wall from the to-and-fro of the outside world. Like the wall of the old mediæval city, the membrane keeps out unwanted entities and lets in important ones.

What are the citizens of the cell-city? There is, of course, water. Dissolved in the water there are ions and salts, calcium, magnesium, zinc, potassium and others that play key roles, for instance acting as chemical signals and triggers to control the cell's activity. But the cell also contains many more complex molecules, typically comprising hundreds to thousands of atoms. These giant molecules include proteins, enzymes, DNA and its cousin RNA, and molecular motors. These giants range from a few nanometres across to a thousandth of a millimetre.

In other words, the cell is full of middle-world-sized objects suspended in water. In the cell, Brownian motion is everywhere[2]. The inevitable restlessness has huge and unavoidable consequences for the basic mechanisms of life.

But first a word about something a bit less grand.

Plastic.

---

2 Some cells, such as bacteria, and other not-quite-alive objects like viruses, are themselves small enough to be middle-world-sized objects.

Chapter 10
THE MIDDLE WORLD CHAIN GANG

Plastic has had a bad press. It has become synonymous with fake – laminate instead of real wood, Tupperware® instead of china, vinyl instead of leather. But like it or not, plastic is a vitally important material. It is the stuff of the 20th century. Bakelite brought design to the masses, shellac and vinyl brought music into the home, and nylon allowed ordinary housewives to indulge in a bit of sartorial elegance.

You can do things with plastic that would be far more difficult and dangerous with materials like metals. You can melt it and make it flow into moulds at low temperatures. You can roll and pull and squeeze plastic much more easily than metals. You can expand it into a light but strong foam or you can freeze it into a solid. You can make it tough enough for bullet-proof vests. It's light and easily transportable and, despite its indestructible reputation, it is actually relatively easy to recycle and reuse.

What's more, plastics have a surprising amount in common with life's molecules. DNA, proteins and plastics are all molecular chains, or polymers[1]. All the miraculous versatility of polymers comes down to the fact that they are inhabitants of the middle world.

Chemistry discovers the middle world
In the early 1920s chemists still thought of molecules as small – assemblages of at most a few dozen atoms. These atoms were so

---

1 Polymer simply means a multi-part molecule: many *mono*mers add up to make a *poly*mer.

closely gathered together – their atomic 'bonds' stretching only to short distances – that the molecules were hardly much bigger than the atoms themselves. Certainly they were much smaller than the Brownian particles measured so assiduously by Jean Perrin barely a decade before: molecular chemistry and the middle world appeared to have nothing to do with each other. Anyway, the quantum physicists were busy constructing maths that would explain small-molecule chemistry in terms of electron orbits and quantum energy transitions. Middle world chemistry was not even a gleam in the eye of the most imaginative quantum mechanic.

A German chemist, Hermann Staudinger, in the mid-1920s, first proposed that molecules could be a lot bigger. Staudinger's idea came from an unlikely source. Every chemist knew that once in a while their reactions tended to go a bit wrong, leaving a gummy sludge at the bottom of the reaction vessel. This sludge, he suggested, could actually be a mass of long chains. Giant molecules – the product of chemistry out of control.

Chemists knew that carbon atoms liked to make ring-shaped molecules: benzene, for instance, is a ring of six carbons. In reactions short carbon chains formed, then looped around and closed on themselves to make small rings.

What might happen, Staudinger asked, if as the reaction proceeded the growing chain missed the chance to loop around to make a ring, and instead just carried on growing longer? Carbon after carbon after carbon... very soon the chain would be so long that it would be impossible to form a stable ring. Such a 'polymer' could have any number of atoms – millions, even.

Suddenly chemistry was no longer just the preserve of the atomic world. With these chains or polymers, chemistry was pushing into the realm of the middle world.

Carbon is particularly good at making chains because a carbon atom can make four bonds with other atoms. Two bonds can connect with other carbons to form a link in the polymer chain, while the other two connect to other atoms such as hydrogen, oxygen or nitrogen. This makes carbon chemically very versatile. It can make anything from a small molecule such as methane (one carbon plus four hydrogens, or CH_4), to shortish polymers that might have a few tens of carbons in a chain (say $C_{20}H_{42}$, a chain of 20 carbons each with its spare two arms attached to hydrogens, and an extra hydrogen at each end of the chain, making 42); to huge polymers such as the plastic polystyrene, with numbers of atoms into the millions.

It took Hermann Staudinger ten years to get the chemistry establishment to accept his idea. Indeed, it wasn't until 1953 that he was awarded a Nobel Prize for it. By then technology had overtaken science anyway. Chemical engineers were already happily making and using long-chain molecules, whether the pure chemists believed in them or not. The engineers were making plastic.

Plastic engineering in the middle world

Much of the versatility of plastics arises directly from their long chainlike nature. The earliest theories of polymeric molecules saw them as rigid: long straight rods like sticks of dried molecular spaghetti. But a polymer is big enough to be subject to Brownian motion, bombarded by other molecules around it, so rigidity is out. Polymers are wriggling, flexible strings.

Polymers do things that small molecules and rigid sticks don't. As they flex and writhe they get hopelessly tangled. Anyone who has tried to sort out the cabling behind their computer or hi-fi will be aware of how easily long flexible chains get tangled up. This affects how plastics flow: as greater force is applied, flow can get easier (because the chains untangle somewhat) or harder (because the flow makes the chains get even more tangled).

The peculiar flow properties of plastics, not seen in simple small-molecule liquids such as water, are put to clever uses in a vast range of everyday products. Non-drip paint is one example. Polymers are added to the paint (a solution of pigment particles in water) which allow it to flow easily when a large force is applied (when you brush it onto the wall), but help to resist small forces such as gravity (to stop it dripping down the wall after application). The polymers respond to the large brushing force by untangling; the small gravity force isn't enough to untangle them, so they hold the paint in place, stopping drips.

Foods such as yoghurts also have polymeric chains added to them to engineer the way they flow (both on the spoon and in your mouth). Manufacturers add polymers to adjust the formulations of their products, according to the verdict of tester panels – foods by focus group. In foods, of course, the polymers aren't inedible plastics. Typically additives are substances derived from cellulose, a long-chain polymer that is a basic component of many plants.

The properties of shampoo are similarly engineered using long polymeric chains[2]. Polymers thicken the shampoo so it doesn't just run off your hair before you get a chance to build up a lather. Long chains also attach tenaciously to hairs, making lasting 'conditioning' easier.

Forces can easily change not just the tangling of polymer chains but their shape too, something that would be far more difficult with small molecules. When stretching a polymeric material like rubber or one of its synthetic cousins, the flexible coiled chains are relatively easy to pull straight. Straightening the wriggling chain is not a question of breaking strong chemical bonds, just of realigning the polymer – so everyday forces can achieve it. On the other hand, the chains don't want to *stay* straight because this cuts down their freedom to wriggle[3]. So when you let go, the chains go right back to their coiled shapes – the rubber elastically returns to its original length.

---

2 Certain types of chain molecule in shampoo are known as worms. Shampoo companies never seem to use that word in their marketing campaigns…

3 In the language of thermodynamics a tangled, wriggling, disordered chain has higher entropy than a straight chain – there's only one way to make a straight chain. So left to itself the chain's most likely state is to be coiled up.

Solid plastics melt at much lower temperatures than metals. The large size and flexibility of the polymer chains mean that unlike metals they don't pack into tightly bound atomic-scale crystals, meaning that manufacturing everything from electronic components to car parts to toys to packaging is easier, cheaper and safer. The middle world makes our lives a lot easier.

This kind of middle world chain technology has been ubiquitous since the 1940s. Before the shampoo reaches your bathroom shelf or the yoghurt ends up on your breakfast table, their formulations will have been puzzled over by many an R&D engineer. Long tangled chains of four-armed carbon atoms are everywhere. Plastics, cosmetics, foods, fibres in clothing, non-stick coatings and non-drip paints, stretchy rubber bands, polystyrene packaging, Perspex windows – and of course nylon stockings. Though many scientists might never have noticed the middle world through most of the 20th century, a vast population of engineers has been working there all along.

Polymer maths and random walks

So much for technology: what about theories of polymers? In the 1950s chemists began to realize that there was a direct mathematical connection between the flexible, wriggling polymer chain and dancing pollen – not to mention the Paris stock market – namely, the random walk. Chemists realized that the idea of the random walk, a series of hops in arbitrary directions, could be recycled and applied to a flexible polymer chain.

This idea was largely the inspiration of an American chemist called Paul Flory. In the 1940s Flory worked originally for Dupont in the USA, under Wallace Carothers, famous for his invention of nylon, one of the first synthetic, i.e. plastic, textiles.

Flory came up with a rather abstract picture of the polymer molecule. He imagined the polymer as a sequence of links, like a bicycle chain. Each link is strong – a chemical bond between carbon atoms – so the chain can't break. But at each link it *can* twist and bend, in a random direction. Going from link to link,

Flory pointed out, is just like taking one of the randomly directed hops made by a Brownian particle. A walk along the polymer, stepping from link to link, would trace out a random walk like the path of a Brownian particle.

Suddenly the statistical maths necessary to understand the behaviour of polymers was revealed, ready-made and ready to go.

Flory also thought hard about what determines the overall shape of the polymer. Would it spend most of its time as a long tangly string, or would it coil up into a ball? The key, Flory realized, was how the links interacted with their environment. In many practical situations a polymer will be immersed in a solvent such as water (this is the case for the chains in shampoo and yoghurt for instance). Flory pointed out that the polymer shape – long open string or tightly wound ball – must be determined by a competition between the way the links interact with the solvent molecules and with other chain links.

If there is, say, a repulsive chemical interaction between the polymer links and the solvent, then the chain prefers to coil up, to minimize its exposure to the solvent. If, on the other hand, the solvent molecules and the chain links attract each other, the chain relaxes and unwinds – swells up, effectively – so as to intermingle as much as possible with solvent. Chemical interactions determine whether the chain is an unsociable, introverted wallflower or a gregarious, enthusiastic mingler.

Flory showed that the simple random walk case represented a unique balance point, where polymer links and solvent had exactly the same interactions with each other as with themselves. At this balance point, a link in the chain can't tell the difference between another link and a solvent molecule. It is equally attracted to (or repelled from) both.

Because the strength of the interactions between polymer and solvent depends on temperature, how close a polymer is to Flory's ideal balance-point, where all interactions are equal and essentially cancelled out, can be controlled simply by adjusting temperature. This major theoretical breakthrough made intelligent polymer engineering far more straightforward. In experiments and manufacturing processes, one could now use temperature to manipulate polymer shape and product qualities – to control whether you got a long dangly string, a fluffy ball or a tightly bundled coiled lump.

For his fundamental work on polymers, Paul Flory won the 1974 chemistry Nobel Prize. At last plastics took their rightful place in the science as well as the technology of the middle world.

At the same time as plastics engineers were making ever cleverer use of middle world polymer chemistry, other scientists were beginning to look more closely at an altogether more complex type of carbon chain molecule: proteins.

Polymers of life

Unlike most plastics – chains of carbon atoms – the 'units' of a protein chain are themselves molecules, complicated chemicals called amino acids. Proteins can be composed of thousands of amino acids. The chemical interactions of these chain units are far more complex than those between single carbon atoms, so a protein represents a new world of chemical complexity compared to plastic.

The first descriptions of the structure of proteins emerged from the Cavendish Laboratory in Cambridge – the same place as Watson and Crick worked out the DNA double helix. One of the major scientists involved was Crick and Watson's immediate boss, Max Perutz, who devoted his whole career to studying the blood protein haemoglobin. Another was Perutz's colleague, John Kendrew, who looked at the muscle analogue of haemoglobin, myoglobin, the protein that stores and carries oxygen in muscle cells.

What Crick and Watson did for DNA, Perutz and Kendrew did for proteins. They measured the first reliable structures of these 'superplastics', these polymers of life.

Max Perutz, like Ludwig Boltzmann, grew up in Vienna. Fortunately for science, he managed to convince his parents that he'd much rather study chemistry than their first preference, law. Inspired by the idea that you could measure the layout of the atoms in a substance by firing X-rays at it, Perutz left Austria in 1936 to do his PhD at the Cavendish.

The head of the X-ray structure team at that time was J. D. Bernal, controversial Marxist, free-love enthusiast and genius *manqué*. As an undergraduate Bernal belonged to the 'Apostles', an exclusive group of intellectuals that also included Kim Philby, later revealed as the KGB's most infamous mole in the British Secret Service. Bernal's nickname at Cambridge was 'Sage', because he 'knew everything'.

Nevertheless he gained a reputation as an underachiever: he would flit from idea to idea but never quite carry things through. 'Sage' Bernal played a key role in building the basic science of structural biology, being the first to make proper X-ray studies of amino acids, steroids, proteins and viruses. As well as Perutz his students included Dorothy Crowfoot, who in 1935 measured the structure of the hormone insulin, for the first time being purified in large quantities for the treatment of diabetes.

In the 1930s Bernal was the first to try to measure the structure of a protein using X-ray scattering. The results were not encouraging: unsurprisingly, since proteins are significantly more complex than the DNA that Crick and Watson concentrated on later. That didn't stop Bernal's new student, Max Perutz, from persevering. Little did Perutz know as he commenced his doctorate in 1937 that the quest for a complete atomic protein structure would occupy the next 30 years of his life.

Slowly the structure-measuring techniques were refined, as was the biochemical knowledge required to obtain pure samples of proteins. Perutz found that he could get good results from samples of horse haemoglobin, which, like human haemoglobin, collects oxygen in the lungs and distributes it via the blood to the rest of the body.

Why horse haemoglobin? At that time it was just a matter of finding the best candidate for success. Proteins are so atomically complex, not to mention difficult to prepare adequately for X-ray experiments[4], that no one was thinking much about *which* protein. In 1937 the largest molecule to date that had been completely pinned down was the pigment photocyanine. It has 58 atoms. Horse haemoglobin has 11,000.

Perutz's colleague John Kendrew meanwhile tried to crack myoglobin, a protein that delivers oxygen to the muscles. Myoglobin has some claim to being one of the simplest of proteins – or so many scientists thought until recent results, as we shall see later, rather shattered that complacency. Kendrew had only got interested in proteins after talking to Bernal while both were serving in Burma during the Second World War[5].

After the war, through the late 1940s and into the 1950s, the X-ray team at the Cavendish and then at the newly conceived Medical Research Institute started by Perutz, spent many person-years refining their experimental techniques. This wasn't quick science or the startling product of an accidental experimental stumble – this was tenacious hard graft. For a long time most onlookers regarded Perutz's and Kendrew's quest as almost hopeless. Sir Lawrence Bragg, Perutz's boss once 'Sage' Bernal moved to the University of London, admitted privately that he thought Perutz's chances of finding the structure of haemoglobin 'indistinguishable from zero'.

Max Perutz and John Kendrew were finally rewarded, in the same year as Crick, Watson and Wilkins, with a shared Nobel Prize for their pioneering work on measuring the structure of

---

4   As in the case of DNA, measuring protein structure by X-ray scattering involves the fiendish task of making crystals of the protein, regularly arranged copies whose X-ray signatures can add up strongly enough to give a discernible measurement.

5   Max Perutz, as an Austrian, was considered an 'undesirable alien' by the British authorities during the Second World War. In 1940 he was shipped off to Canada. The authorities relented and let Perutz return to Britain a year later; he even ended up working on secret war plans to use icebergs as aircraft carriers.

proteins. This was in 1962, almost thirty years after Perutz's first experiments. Even then the job wasn't finished. A precise structure for haemoglobin was not actually published until 1968.

One of the key findings of all this work was that (surprise, surprise) proteins are indeed a big step up in complexity from plastic. Protein structure turns out to be far more complex than the simple random walk of a plastic chain. Each of the amino acids in the protein is already a complicated chemical: so the links in the protein chain have a fantastically complicated set of interactions. The balance of those interactions against interactions with the protein's environment leads to a vast structural complexity.

It is a vitally important complexity. As scientists slowly learnt over the years following those first structural measurements, this complexity is responsible for the ability of proteins to carry out such a wide range of precise chemical functions in living things. We very much owe our existence to this balance of interactions in protein chains. Without it, proteins would be useless tangled lumps of atoms, no more capable of supporting life than a lump of plastic.

But still, there's something missing. As if the chemistry wasn't already complicated enough, there's also the jiggling. A protein chain, just like a polymer, is a middle-world-sized object surrounded, in the cell, by water. Proteins can't stand still. It's not just chemistry that's important, but motion too.

Dynamic proteins: platonic vs. 'kicking and screaming'

Proteins do most of the necessary tasks in the cell. They are the motors that ferry chemical cargo around, the enzymes that catalyze vital chemical reactions, the machines that copy DNA, the pumps that control traffic in and out through the cell membrane, and so on. Proteins even make other proteins by reading and transcribing the genetic code.

They are the machines of life. They must move to function – and, being middle-worlders, they must move anyway. The modern challenge in protein science is not just shape – but motion. As biochemist Gregorio Weber aptly wrote in 1975, proteins are not the 'platonic' molecules beloved of X-ray structural studies – they are 'kicking and screaming' chemical beasts.

One of the pioneers of protein motion experiments was American Hans Frauenfelder. It was at a conference on measuring the structure of biological molecules in the late 1960s that Frauenfelder first got interested. He went around asking delegates the following question: what about going beyond protein structure – and measuring motion too?

Forget it, they said. That's just too difficult a problem.

Frauenfelder's curiosity was aroused, and he returned to work determined to glimpse, somehow or other, a protein in motion. Like Kendrew, Frauenfelder chose myoglobin as the simplest candidate for his experiments. At the time it was impossible to see the motion of any part of a protein directly. The advantage of myoglobin was that, buried in its centre, is an iron atom. This iron is central to myoglobin's function. Oxygen molecules bind to the iron, to be carried and released when and where necessary. Frauenfelder realized that measuring how oxygen molecules made their way into the protein to reach the iron buried in the middle was an indirect route to measuring the way the protein structure itself moved and shifted.

Some kind of fluctuation and shape-change of the protein structure was clearly vital to its function. Because, paradoxically, existing measurements of myoglobin structure indicated that the iron atom was completely buried by the twists and coils of the chain. For molecules such as oxygen there was just no way in.

The one possibility for the oxygen to make it into the protein and reach the iron, was if the parts of the chain shifted continually around, opening up temporary oxygen-sized gateways. The middle world dance of the myoglobin molecule was the secret to the way it worked.

The concept of Frauenfelder's first experiment was straightforward. Given a myoglobin with its cargo in place, force that cargo molecule to detach from the protein. Then watch how the cargo moves as it finds the iron again and reattaches. The process must report indirectly on how the protein shifts around as the cargo 'searches' for the place to hook itself back in.

To detach the cargo from the protein, Frauenfelder's team used a technique called flash photolysis. Energy from a laser flash, applied at the right frequency, snaps the chemical bond between the iron and the cargo molecule.

The cargo chosen by Frauenfelder was not oxygen, myoglobin's normal biological payload, but carbon monoxide. This was a better candidate for the measurements because under normal conditions oxygen can bind and unbind all by itself, so confusing the results. Once carbon monoxide is bound, only the laser flash can dislodge it. (Carbon monoxide's unwillingness to unbind from myoglobin and haemoglobin is the reason why gas leaks can be fatal: carbon monoxide displaces the oxygen in the blood and tissues, and then can't be removed. The oxygen supply system breaks down and cells run out of energy and die – the victim is asphyxiated.)

By the early 1970s Frauenfelder and his team of students had their experiment up and running. What they found unveiled an unsuspected web of complexity in the internal motions of this apparently 'simplest' of proteins. A myoglobin protein, they discovered, has a whole host of slightly different conformations – different shapes, different layouts of the atoms in space. The protein continually swaps and re-swaps between these conformations, wriggling and rattling from shape to shape.

Moreover by comparing the reaction behaviour of given proteins in various of their conformations, Frauenfelder's team and others showed that the precise shape of the protein has important effects on its function. What the protein can do and how well it does it depend sensitively on the detailed shape of the chain. As recently as 2001 an extreme example of this 'shape-determines-function' effect was discovered, when myoglobin was shown to have an alternative function, quite separate from its usual employment as oxygen carrier-bag. Under acid conditions myoglobin changes shape and becomes a catalyst for a chemical reaction producing nitrite ions instead. Even the textbook example of a simple protein turns out to be a multi-purpose molecule whose function is determined by exactly how its shape responds to its environment.

Frauenfelder's team showed that some of the fluctuations of the myoglobin protein are directly driven by the molecular bombardment of the surrounding water. The team essentially captured a direct view of Brownian motion, now for a complex functional molecule rather than a pollen particle or a sphere of *gamboge*.

The experimenters got a direct view of how fluctuations enable myoglobin to work – the solution to the no-way-in-for-oxygen paradox. As parts of the chain wriggle under the bombardment of water molecules, the protein continually shifts between its possible conformations. Some of these conformations include momentarily open gateways that cargo molecules such as oxygen can squeeze through and reach the iron buried inside.

The function of the protein turns out to be a fantastically complex balance of chemical choreography and Brownian chance. Indeed, this balance between chemical rules and the randomness provided by Brownian motion is perhaps *the* fundamental theme of middle world chemistry and biology.

After three decades studying protein motion, Hans Frauenfelder is now a sprightly, if somewhat wizened, emeritus member of the Los Alamos laboratories in New Mexico. Even now, to his and other protein dynamicists' despair, biochemistry textbooks often give an all too staid impression of molecular biology as static sculpture rather than dynamic whirlwind. Frauenfelder has experienced first-hand, over 30 years of research, how fundamental protein motion is to protein function.

Fluctuations amongst working conformations are only part of the protein story. Perhaps the greatest mystery investigated over the last few decades is how a protein finds a useful conformation in the first place. How does a complex middle world chain, manufactured in the form of a loose, gangly string of a few thousand atoms, ever shape itself into an efficient tool of the cell? This is the so-called 'folding problem': the Holy Grail of protein science, with middle world restlessness at its heart.

Molecular origami in the middle world

A protein is a long chain of amino acids. But a *working* protein is a very particular shape, a precisely defined coiled structure that enables it to do a specific job in the cell. The way proteins take up their precise shapes, starting from long straight chains of amino acids, is known as protein folding: the molecular origami by which the bits of the chain end up arranged in just the right way.

A prime example of the importance of protein shape is the class of proteins known as enzymes. Enzymes are catalysts, helper proteins that improve the efficiency of chemical reactions. They work, basically, by making sure that everything is in the right place at the right time.

That's what chemical reactions are all about. In a chemistry lab (or in the kitchen) you can take your reactants (ingredients), put them together in a test tube (mixing bowl), and shake (give it a good stir) to bring everything into contact, so it can react (make cake).

In the cell, molecules have to come together under their own steam – there's no one to put them in the right places. More than that, reactants need to come together in just the right way, with their reacting bits just beside each other: only with such efficiency can a cell survive in a difficult and competitive environment.

So most biological reactions demand a precise spatial docking so that reacting components lock like jigsaw pieces, and stay locked for long enough that the exchange of energy and matter can occur with maximum ease. Like two porcupines making love, if you don't get the configuration exactly right, no matter how keen the participants, it just isn't going to happen.

To ensure this efficient docking of chemical porcupines, it is not enough to let nature take its course. In the cell enzymes take up the task of shepherding the right chemicals together in the right way, so they can 'get it on'. To draw in the right chemicals and hold them in just the right positions to encourage a reaction an enzyme protein needs a very particular shape.

In the 1950s American chemist Christian Anfinsen began pioneering experiments on protein folding. He showed that proteins had a mysterious memory. A typical protein, Anfinsen found, even when turned into a long shapeless chain by being heated or put in an unfamiliar environment, would very quickly spring back to the correct folded shape once returned to its natural habitat.

How? The most obvious answer is that the shape is dictated by the interactions between the protein's components. Driven by those interactions the chain coils up into its most stable conformation. Again there's something missing, however.

Chemistry is not enough: how should the protein's atoms *find* their most comfortable conformation unless they can try out all the possibilities? You can't get comfy on the sofa without wriggling around a bit to find the best place. So a protein needs to fluctuate, twist and writhe until it finds the most stable folded up structure, the one that best balances the chemical interactions between its component atoms and the surrounding water.

Of course the middle world dance is fundamental to this wriggling folding process – it *is* this wriggling process, this search for the most comfortable shape. Middle world motion is fundamental to how the protein even gets to that structure, not to mention how it works when it gets there.

Even once middle world fluctuations were taken into account, a peculiar mystery lay in wait for the proponents of the 'search for perfect comfort' theory of protein folding and protein shape. In the 1960s American chemist Cyrus Levinthal pointed out the problem, which came to be called the Levinthal Paradox.

Levinthal demonstrated that an unfolded protein of even a few thousand atoms – and many proteins are far larger than this – does not have *time* to try out all possible shapes to find its most stable one. Measurements by Christian Anfinsen and others revealed that a protein can fold into its stable shape within a fraction of a second. Even given very generous estimates of the rate at which a protein might be able to try out various shapes, Levinthal showed that it would take a protein much longer than a fraction of a second to find its minimum energy configuration if it were testing all the possible arrangements of a chain of thousands of atoms.

So there was a puzzle buried inside the origami of proteins. It was almost as if the protein had a pre-knowledge of what sort of shape it should be. Chemistry was once again not enough. The key to the folding problem was to understand more about exactly how a protein explores the middle world.

The lie of the land

Our picture of how proteins wriggle, fluctuate and fold has advanced enormously over the past few decades, as a result of experimental advances and because of leaps in computer power. Calculations of protein structure can now take account of interactions almost atom by atom. More and more scientists have been getting in on the act pioneered by Hans Frauenfelder, searching for insights into how proteins hop around from conformation to conformation.

Three decades of work on protein folding has resulted in a concept known as the energy landscape of the protein. This is essentially the complete universe of conformations that a protein can explore by swapping between shapes.

The energy landscape is like a mountain range. Valleys represent the energies associated with particular shapes, and the heights of mountains between valleys represent the difficulty of swapping between those conformations – the energy cost of such a swap. The likelihood such a shift in shape is determined by that energy cost – according to the statistical rules first worked out, for the simpler case of atoms in a gas, by Ludwig Boltzmann.

The energy landscape that a protein explores turns out to be marvellously complex. For instance Hans Frauenfelder's myoglobin results indicate a hierarchy of protein superstructure:

in the language of the landscape, valleys within valleys within valleys, mountains upon mountains upon mountains.

Most of the time the protein's shape fluctuates by small amounts, easy low-energy-cost hops that explore a local area of the landscape – one valley, say. But just occasionally, as it probes the landscape, the protein stumbles on an easy gateway into another distinct zone – a low-energy-cost pass into a neighbouring valley. Jumping into this new region allows it a big switch into a quite different set of conformations and functions. (Myogoblin's shape-change between oxygen carrier and nitrite reaction cata-lyzer is an example of such a switch.)

Consider the energy landscape as a map of Europe – a conti-nent with its many connected yet distinct countries, each with their own cultures. Say you are born in the UK and you spend your early life there. Though your insatiable fidgety desire for travel sees you hopping around the country, nevertheless these are local hops – the immediate environment stays more or less familiar. Everywhere there is warm beer and fish and chips. One day a momentary gateway opens into another country – you win some Eurostar tickets and pop across the Channel into France. Now your local environment is different – with your small fidgety hops you explore a new land characterized by fine wines and excessive consumption of cheese and garlic. In the 'valleys and mountains' picture you have jumped from one valley – the UK – to another – France.

Later your unavoidably itinerant lifestyle might happen to carry you close to the German border. You step across – and sud-denly you find yourself in yet another new zone, this time with all the distinct local characteristics of Germany – efficient cars, trains that run on time, an obsession with sausages. But who knows how long you'll stay there – perhaps a chance hop south-wards will soon see you in Switzerland, amongst the chocolate and cuckoo clocks...

Most of the time your hopping about is local, and the local landscape stays familiar. Now and then, when you find a low-cost route to another country, the environment changes substantially. So it is for the protein in its complex energy landscape: many rapid small shape changes overlaid on occasional big shifts to quite different conformations.

All this is dictated by the combination of the chemical interactions between the protein's component atoms – setting up the energy landscape – and the inevitable dance of the middle world that provides the power to explore.

The energy landscape idea offers at least a partial answer to the Levinthal Paradox: how a protein can find its most comfortable shape so quickly. Broad features of the landscape are determined by broad features of the protein structure, such as the major links in the amino acid chain. This acts as a sort of pre-programmed underlying landscape that is still there even when the protein is unfolded, as in Christian Anfinsen's experiments. On starting to refold, a protein already has this underlying contour map built in. So refolding is more a question of quicker local wriggles than a complete re-exploration of the whole map.

Nevertheless, the folding problem remains far from fully understood. Scientists run regular competitions, for instance, to see who can calculate the best folded protein structures given only the sequence of amino acids in the protein: how close can you get to the real experimental structure on the basis of chemistry alone? Somehow – with the vital help of the middle world – nature still does this very much faster and more reliably than scientists can, even with their monster computers.

The state of the art in wriggling proteins

Hans Frauenfelder's earliest experiments on the motion of carbon monoxide molecules inside myoglobin were an indirect route to measuring protein motion – a sort of sideways glance. Recently, experiments have more directly illuminated the protein dance quite spectacularly.

In 2005 a team led by Giancarlo Baldini at the University of Milan-Bicocca in Italy published a study of the structural fluctuations of a naturally fluorescent protein called Green Fluorescent Protein or GFP. GFP is derived from the luminescent jellyfish, *aequorea victoria*. GFP's usefulness in protein dynamics experiments is that it fluoresces in *different* colours, green and blue, depending on which of its two main shapes it is in. Using a laser to induce the fluorescence, Baldini watched in the microscope as the protein wriggled between conformations – changing from green to blue to green again like some protein traffic light. Trapping the protein in a tube of silica gel, the team watched a single protein molecule for many hours, in a watery environment similar to the cell. Instead of obtaining averaged results over many proteins as they wandered in and out of the microscope's field of view, the team tracked the behaviour of just one molecule[6].

The team looked at the protein *unfolding* – the reverse of the original folding problem – because GFP can be chemically induced to unravel in a controlled way. In its unfolded, floppy-string state the protein is not fluorescent. Hence unfolding is visible as a 'disappearance' of the glowing molecule.

What Baldini and his colleagues saw as the GFP unfolded surprised them. Instead of just switching from one folded configuration (green, say) straight into the floppy unfolded string (i.e. dark), the molecule started to flicker in a complex way between the green and blue states as the moment of unfolding approached. At first the shape-shifting seemed random; then as the moment of unfolding came closer the protein oscillated regularly between blue and green, quite as if it was gearing itself up for the big unwrap. In the last few milliseconds before unfolding, the shape fluctuations lost their regularity and became wilder than

---

6   Copies of complex chain molecules – plastics and proteins – can behave subtly differently even when chemically identical, a phenomenon christened 'molecular individualism'. At its root lies the sensitive randomness of the middle world, as we will see.

ever. Finally, as unfolding took over, the protein ceased to be fluorescent and disappeared from view.

In a final twist, Baldini's team found that a given molecule, if allowed to refold and then forced to unwrap again, did not behave the same second time around. On its next unfold it might oscillate between green and blue at a slightly different frequency. Subtle differences in the way the protein was refolded seem to have significant consequences.

Similar results from a host of other experimental methods are beginning to pile up. Again in 2005, a team headed by Dorothee Kern at Brandeis University in the USA used a method called nuclear magnetic resonance (NMR) to study the shape-shifting of the enzyme cyclophilin A. Kern and the team found that, even when not at work helping a reaction – during a 'tea break' in other words – the enzyme constantly fluctuates between conformations, wriggling around in response to the bombardment of surrounding water molecules. No surprise there, although it was nice to see the middle world dance again in the flesh. Interestingly, these 'tea break' fluctuations happen at rates very similar to those seen when the enzyme is actually at work. The Kern team's interpretation is that these apparently idle 'off-duty' shape digressions are still fundamentally part of how the enzyme works when it is actively on the job.

As protein experts Yuanpeng Huang and Gaetano Montelione wrote in a commentary on the Kern team's results in the journal *Nature*:

> Static pictures of protein structures are so prevalent that it is easy to forget they are dynamic molecular machines... static protein structure is only part of the story of protein function; the other half tells of mobility, and many chapters are left to be written on how the intrinsic dynamics of a protein's structure can provide an underlying basis for its biological functions.

In other words, after many years admiring static protein sculptures, protein scientists are now being dragged 'kicking and screaming' into the middle world, whether they like it or not. And about time too: as history has shown since 1827, the middle world is a surprising, exciting, vital place – a great place to explore.

A dangerous balancing act

The very sensitivity of protein function and dynamics to the precise chain shape can have grave consequences. Accidental mistakes in protein folding are thought to be responsible for some diseases, such as Alzheimer's and the human form of mad cow disease, Creutzfeldt–Jakob Disease or CJD. CJD is thought to involve a misfolded protein called a prion. Incorrect folding means the protein cannot do its job properly. In CJD, moreover, somehow just one misfolded protein seems to encourage others to misfold too, resulting in a wholesale breakdown of operations in the nervous system and the brain.

Perhaps the real lesson here is that life is a result of a terribly fine balance between the rules of chemistry and the randomness of the middle world. Without randomness, all would be over and done in a quick flash of chemistry: there would be no dynamics, no change, no life. Too much randomness, and life's processes lose control, resulting in chaos.

Those efficient chemical reactions, efficient ways to harvest the chemical and energetic resources around us, can only be achieved by the most precise balance of rules and randomness. Too many rules – or too much randomness – and the whole house of cards collapses.

Half a century of work on the structure and motion of proteins has shown that proteins are at the heart of the rules/randomness balance in the cell. There is yet one particular set of proteins that deserve a chapter on their own. These proteins use this fine balance of chemistry and Brownian motion to fulfil one of the principal requirements of life. They enable organized cellular life to get going.

These proteins' job is to turn energy into useful work. They are the engines of life.

Chapter 11
ENGINES OF LIFE

Something astounding happened in the middle world about four billion years ago. Something that had strange echoes rather more recently – in 19th century human society.

In the 19th century, scientists, engineers and industrialists finally understood the rules of converting energy into work. Ironically, unbeknownst to these human pioneers, people like Sadi Carnot, William Thomson and Rudolf Clausius, the energy-into-work problem had actually already been solved. Someone – or rather something – had got there first. Something rather close to Carnot and the others – something, in fact, deep inside each of them. And it hadn't just pipped them to the post in the race to convert energy into work: it had won the contest by a clear margin of almost four billion years.

That something was life.

About four billion years ago, somewhere, somehow, there was a Bioindustrial Revolution. It was the moment when the engines of life appeared, when life discovered how to convert energy into work. When the industry of life really started to harvest the natural resources of the Earth.

The Industrial Revolution that reached its peak in the 19th century may have changed human society – its long-term consequences may even destroy human society. But the Bioindustrial Revolution did something far more substantial. It changed the entire planet from a chemical rock to a living breathing globe.

Scientists today are learning about the Bioindustrial Revolution in unprecedented detail by studying a class of proteins – the molecular engines of life. But to see how they are doing that, first we have to take an unlikely sounding detour… into a world of very cold atoms.

B is for bacterium; A is for atom-trap

In 1997, the Nobel Prize for Physics was won by a triumvirate of scientists: in Paris, Claude Cohen-Tannoudji of the College de France; and in the USA, Bill Phillips of the National Institute of Standards and Stephen Chu of Stanford. The Prize citation reads: 'for development of methods to cool and trap atoms with laser light'. This is a prime example of how you can never quite predict where scientific research is going to lead. Tannoudji, Phillips and Chu spent years thinking about and investigating the quantum properties of atoms and photons, and inadvertently opened up the middle world of cellular life.

In 1985, Stephen Chu demonstrated that a pool of tightly focused intense light from a laser beam can be used as a sort of honey-jar for atoms, slowing them down so that instead of zooming through the almost empty air at hundreds of centimetres per second they wade through a viscous 'optical molasses' at almost no velocity at all. Continual interactions with photons of laser light steal the atoms' energy. The result is a small blob of atoms trapped in the focus of the laser beam.

This sounds as about as far as you can get from the middle world or engines of life. That would be reckoning without one of Chu's colleagues, Art Ashkin, who had the office next door to Chu's at Bell Labs when Chu arrived there in the early 1980s. Back then, Chu was thinking not so much about slowing atoms as about how to achieve very low temperatures – by cooling materials with laser light. *Cooling* something with a laser sounds counter-intuitive. Recall the famous scene in *Goldfinger*, where a laser hot enough to cut metal threatens to vaporize James Bond's wedding tackle. But theory implied that, arranged in the right way, colliding photons of light could be made not to add heat but to subtract kinetic energy from atoms, and so effectively cool them down.

Ashkin's interest in the way atoms interact with light went back to experiments he had performed at the end of the 1960s. Ashkin had done a rough calculation, out of curiosity, of the force that light ought to be able to exert on a small particle, say a bead one millionth of a metre across. Encouraged that this force, according to his calculation, ought to have visible consequences – be strong enough to move the bead around – Ashkin set up the experiment using a laser to provide an intense source of light. He shone the laser into a beaker of water containing suitably small beads of polystyrene. Ashkin's experiment has echoes of Robert Brown's: both of them looking down a microscope at tiny dancing beads. Except that when the random dance of one of Ashkin's polystyrene beads brought it into the laser beam, the bead got a visible kick. The force of the intense light was indeed enough to push these small particles around.

Ashkin's polystyrene beads were of course middle-world-sized particles. Still it took a long time for the significance of that to be grasped. Ashkin, like Chu later, was still down in the microworld, thinking about single atoms. If laser force could manipulate polystyrene beads, why shouldn't it be used to grab hold of individual atoms? Such a tool would provide direct control over the elemental units of matter.

Cue Stephen Chu and Art Ashkin running into each other at Bell Labs in New Jersey, in the autumn of 1983. Ashkin was still nursing his atom-grabbing dream, but Bell had actually closed down the project a few years previously, despairing that anything would come of it. Ashkin was searching for a way not just to push atoms around, but to hold them and manoeuvre them with laser light. He'd never managed to work out a way to do this. Suddenly Chu saw the real problem and the way around it.

While laser light should in principle be perfectly able to hold atoms, there was no way to do so for long, because at room temperature they would be moving too fast, at their normal speed of hundreds of metres per second.

The way to hold onto atoms was to cool them down – *slow* them – first. And Chu knew how to do that, again by using lasers. So Ashkin's interest in laser-trapping combined with Chu's interest in laser-cooling led, less than two years later, to Chu's 'optical molasses': a way to slow, grab and hold a blob of cold atoms[1].

But in the meantime, while Chu's team struggled with the technical difficulties of cooling atoms, something at the back of Art Ashkin's mind kept tugging him back to the middle world. Waiting for Chu to perfect the atom-trapper, Ashkin decided to make a 'toy' version of the experiment, once again with beads in water. He realized the trapping should already work with glass beads in water. A bead say a thousandth of a millimetre across, in water, already moves very slowly compared to an atom: the viscosity of the water around it and its size mean that it is already in its own kind of molasses, doing the comparatively stately middle world random walk.

Ashkin called his technique 'optical tweezers' – and very soon its usefulness in investigating the middle world of life became clear. Ashkin started using his tweezers on virus particles – appropriately middle-world-sized rods of the tobacco mosaic virus. While trapping and manipulating these beauties with his lasers, Ashkin noticed there was some other kind of tiny particle in his water beakers that also got trapped by the optical tweezers. Putting the whole setup under the microscope, Ashkin realized that these were bacterial cells – slowly multiplying over time in his beaker, and occasionally wandering into the tweezer beam and getting caught.

By now Ashkin was well and truly into the middle world. Though the lasers used at first tended to zap the poor bacteria

---

1   After the first week-long experimental run, spent busily measuring things with electronic photodetectors and other clever gadgets, it suddenly occurred to someone on Chu's team just to *look* through the window of their trapping-machine. They saw with their own eyes what they had created – a gleaming blob of slow atoms!

('death by opticution' Ashkin termed it), he soon discovered that an infrared laser instead of a normal optical beam could catch the bacterial cells alive. That discovery had huge consequences.

Ashkin realized there was a goldmine of things you could do in cell biology never possible before optical tweezers. Soon researchers were using the optical tweezers to do everything from testing the elasticity of blood cell membranes[2], to separating bacteria one by one in the search for special super-bugs, to testing treatments against influenza viruses by directly bringing treated cells into contact with virus particles. Researchers even optically manipulated live sperm cells during *in vitro* fertilization: test-tube babies by laser light!

It was not long then before scientists hit on the idea of using optical tweezers to study the engines of life.

There is a peculiar postscript to the story of the cold trapped atoms that harks back to the middle world[3]. While Art Ashkin was shuffling bacteria with his optical tweezers, Chu finally got the atom trap working: slowing atoms down to an effective temperature of a thousandth of a degree above absolute zero. Battered back and forth by the sea of laser photons, the slowed atoms' motion turned into a random walk. Brownian motion, but with sodium atoms instead of pollen, and photons of light instead of water molecules[4].

---

2 The elasticity of cell membranes can be affected by diseases such as cancer, so such a measurement can be a useful diagnostic tool.

3 In fact more than one. Chu's maternal great uncle studied physics at the Sorbonne, under none other than Jean Perrin.

4 Chu's team went on to study the spread of atomic velocities – directly measuring what James Clerk Maxwell and Ludwig Boltzmann had predicted with their revolutionary statistical theories more than a century before, at a time when to them the atom was still a completely hypothetical beast. Moreover, by allowing sets of atoms with velocities in a selected range to escape the trap, they manipulated the atom energy distribution, playing with the fundamental statistics of the atomic world.

Plastic individuals: surprises from middle world chemistry

In 1988 Chu, now at Stanford, also began to get interested in using optical tweezers in biological experiments. His first interest was to look at how proteins made copies of DNA, the first step in putting the genetic information to use. But after running into difficulties Chu's team decided instead to look at DNA itself.

DNA's interaction with the tweezer light wasn't strong enough for them to grab it directly. So they attached a 'handle' to the biomolecule, a polystyrene bead not so different from those used by Art Ashkin. By 1990 Chu's group had developed the technique to the point that they could attach a bead-handle to a single, fluorescently labelled DNA strand and move it about in a sea of non-fluorescent but otherwise identical DNA. Under fluorescent light, the single labelled DNA chain lit up against a dark background of non-fluorescent molecules. The team could watch and manipulate their labelled DNA and see how it behaved immersed in a sea of fellow molecules.

This might sound an odd thing to do. A sea of DNA isn't the sort of environment found inside a real cell, for instance. Chu was still thinking more as a physicist than a biologist. He was really trying to use DNA as a 'model' of a polymer.

Just as Jean Perrin needed the 'ideal', simplest possible Brownian particle to study the basics of Brownian motion, Chu was looking for the 'ideal', simplest possible polymer to study the basics of polymer physics. The qualities required were essentially that it should be long, thin and wiggly. Chu's thinking was that DNA, while chemically rather complex, was physically an excellent candidate for investigating basic behaviour, such as how a polymer's wriggling is affected by being immersed in a sea of other wriggling polymers.

Chu's team found that DNA strands from a virus called a lambda-phage were just right for their experiment: big enough to see, grab and handle in the microscope, small enough to represent run-of-the-mill polymers. With the optical tweezing technique Chu held onto a single DNA polymer and set about teasing out the secrets of middle world chain-gang behaviour. His team subjected the polymer to a barrage of experiments, including stretching it and watching it coil up again, and pulling it along to see how it dragged its way through the sea of other polymers.

First, Chu and his team confirmed the basic theory of how a polymer should move in a crowded sea of its fellows: by wriggling in the fluctuating 'tube' formed by its also wriggling neighbours.

Then they stumbled on something quite unexpected. They found that individual molecules can have a personality all their own.

The optical tweezers allowed the group to select individual DNA molecules and feed them one by one into a minuscule pipe through which water flowed like a microscopic river. The flowing water stretched out the DNA molecule. Different DNA molecules, even though chemically identical and subjected to an identical stretching force in the channel, responded in individual ways. Some drew right out; some tied themselves into knots; others remained stubbornly coiled.

One of the tenets of chemistry is that molecules with a given chemical recipe are the same. With small molecules it's true: one molecule of carbon dioxide is indistinguishable from another. Chu's group demonstrated that in the middle world, things are different. Polymers can't be considered identical anymore, even though made of the same sorts of atoms. The continual bombardment by surrounding molecules makes the precise spatial structure of any given polymer at any given moment likely to be different from all its neighbours. Chu's group showed that those tiny differences – an extra bend here, a different twist there – translate into important variations in the way each polymer responds to a force.

Middle world molecules, in other words, are all individuals.

The cargo truck of life

While Stephen Chu and his team were investigating the new world of individual chemistry, other applications of optical tweezers were mushrooming – and getting ever cleverer. When researchers began to use them to measure forces, they were at last on the track of the Bioindustrial Revolution.

Forces are the domain of life's engines. Engine proteins use chemical energy to generate force – converting energy into motion. In this they are in principle no different from the engines of our own Industrial Revolution. A steam engine converts energy from coal into the motion of a piston.

Engines of life, however, turn out to be subtly, yet fundamentally, different from the engines we are more familiar with. At the heart of that difference is the restlessness of the middle world.

One of the first protein engines to be studied in detail with optical tweezers was a molecule called kinesin. Kinesin is the cargo-truck of the cell.

A cell isn't some wet bag stuffed haphazardly with a jumble of giant molecules – proteins, DNA, enzymes and other middle world what-not. It's much more than this: the cell is organized.

Think of a kitchen in a busy restaurant: if staff were trying to chop vegetables, fry steaks, bake bread, slice gateaux, dress salads, do accounts and wash up, all in the same space, the result would be a disaster. There would be vegetables in the gateaux, dishwater in the salad dressing, and lettuce leaves amongst the tax returns. If the kitchen is organized – hotplates here, sinks there, preparation areas over there, the manager upstairs in the office totting up the books – then there's a better chance the restaurant will function efficiently.

Similarly, in the cell everything happens in its place: nutrients are drawn in and waste expunged through specific channels in the membrane; new proteins are made in a special zone called the ribosome; DNA is read and copied in the cell nucleus.

To achieve this organized activity, the cell requires a transport system. At the membrane those nutrients, once drawn in, have to be carried to the right places; to make new proteins in the ribosome requires raw materials to be brought in from elsewhere; those newly minted proteins then need transport out to take up their new employment. And of course, because the cell and its inhabitants are located in the middle world, the transport infra-structure must function there – with all the random disturbance we have come to expect.

A transport network has two essential parts: vehicles and tracks. Take the vehicles first. Kinesin is one of the main cargo carriers in cells. Like any protein, kinesin has certain specific structural qualities for the job. One end – the head – is particu-larly good at bonding to many other molecules: it is the 'truck' part of the vehicle, the place where the cargo attaches. At the other end – the feet – are two pieces that act as the engine and wheels of the truck[5].

The other component of the transport system – the track – is a crisscross network of fibres made from another protein, tubulin. These 'microtubules' form a scaffold network of transport path-ways. Kinesin proteins walk along the tubulin network, delivering their cargo wherever it's needed[6].

The kinesin engine was first identified and named in 1985 by a team of scientists including Ronald Vale and Michael Sheetz of the Woods Hole Marine Biology Laboratory in Massachusetts, USA. They isolated and purified a protein from samples of squid nerve cells. The same protein was also found in cows' brains. Vale named it 'kinesin' from the Greek word 'kinein', meaning 'to move'.

What a kinesin molecule actually does, in the language of our old friend thermodynamics, is convert some kind of stored energy into useful work. It generates forces that move its feet along the intra-cellular highway. Where does the energy come from and

---

5   In the journals, what I've called the 'feet' of kinesin tend to be called the 'heads': I call them 'feet' here because it makes the locomotive purpose of these bits of the protein more obvious.

6   Transport is important beyond raw material supply: for instance 'killer' particles that attack foreign invaders are carried by kinesin along microtubules inside lymphocyte cells. Kinesin also carries sig-nalling molecules, important in brain and nerve cells.

how does the kinesin engine convert it to useful work and motion?

Here's the 'chemistry' explanation – as far as researchers understand it so far, after two decades of non-stop research. In common with many other biological processes, kinesin gets its energy by processing the 'fuel molecule' ATP. ATP molecules waft around the cell like tiny moving petrol stations. A kinesin protein attached to the tubulin track catches an ATP molecule from its surroundings and binds to it. The process of binding alters the kinesin molecule's shape. This shape change, in turn, changes the way the kinesin is attached to the track – effectively supplying energy which enables the kinesin to lift up one of its feet. And that lifted foot shifts forward. (Caveat #1: These last six words hide a vast mystery, as we shall soon see...)

Meanwhile, the ATP molecule stuck to the kinesin has been busily undergoing a chemical reaction, converting into adenosine diphosphate (ADP). ADP and kinesin are much less attractive to each other, so the ADP is released. No longer bound to the fuel molecule, the kinesin switches back to its original shape. This means the lifted foot, 'de-energized', is drawn strongly back to the tubulin track. Having taken a step forward, the foot reattaches. (Caveat #2: Once again, that phrase 'having taken a step forward' is where the problems really start. See below...)

Then the cycle happens all over again, with the *other* foot: binding to ATP causes change of shape; energizing and lifting the foot; foot swings forward; ATP transforms to ADP and is released; causing kinesin to change back to original shape; foot is de-energized and attaches to tubulin again, one step further along the scaffold... and again and again. Doing the ATP-powered shuffle, the kinesin cargo truck gradually makes its way from A to B.

Much of this knowledge of how kinesin works comes from basic chemistry experiments – measuring reaction rates, checking how different levels of ATP-supply affect the motion, and so on. But the optical tweezer ushered in a new type of experimentation, wherein engines like kinesin could actually be seen in motion and manipulated.

For instance, in 1993 a team including Karel Svoboda and Stephen Block of Harvard University decided to measure just what forces the kinesin engine was capable of producing. They attached glass beads to single kinesin molecules, and trapped and tugged the beads with optical tweezers. The tweezer effectively applies a force for the moving kinesin to fight against. Thus the team controlled the load that the kinesin locomotive had to carry, and measured directly how the engine responded to different demands. Pushing the kinesin almost to overload – at a force of about 5 piconewtons, equivalent to lifting about a thousand-billionth of a kilo – Svoboda's team slowed the kinesin enough to see its individual steps – molecular strides of about 8 billionths of a metre.

The past decade has seen this engine of life well and truly put through the wringer. Optical tweezers have been combined with a host of other techniques, such as genetic manipulation to make 'mutant' kinesins with, for instance, only one foot, or no head – all to see just which bits of the molecule are really vital. It is almost as if scientists had discovered an alien technology and were busily subjecting it to a barrage of tests to understand how it works. Except there's nothing alien about kinesin or other protein engines. They're what make our cells tick.

The exact biochemical details of how kinesin functions are still far from well understood. It is an exceptionally complex machine. But never mind the details – there is still one huge problem that overshadows everything the biochemists discover.

Time for those caveats.

The impossible engine

There is a serious gap in the description of the kinesin engine as pictured in the previous section. As a protein, kinesin is a middle world molecule. The 'chemical' description of the kinesin engine is accurate – but it ignores its inevitable restlessness.

Truth is, middle world machines just can't work this way.

Engines need moving parts: the camshaft in the internal combustion engine, the piston in the steam engine. Something has to be able to move in order to provide motive force.

In kinesin's case, to enable motion the protein has to be able to lift up its feet and take steps. But the feet are pieces of protein. Unlike the pistons in your car engine, the moving parts of the kinesin engine are middle world objects.

As soon as kinesin's foot lifts up, it is battered by collisions with the surrounding water molecules. It can only respond by wavering around all over the place in Brownian motion. Brownian motion is inherently random. This means that once the foot is lifted, it ought to be just as likely to step *backwards* as forwards. The cargo-truck seems doomed to wander randomly back and forth along the track, rather than get quickly from A to B.

This is the difference between the engines of the Industrial Revolution and those of the Bioindustrial Revolution: the moving parts of car and steam engines are large, macroworld objects and don't have to deal with random buffeting. They can be made to go in one direction only. The moving parts of bioindustrial engines, by contrast, are slaves to continual fluctuations, infected by randomness. This should make bioengines, and therefore life, impossible.

Oops.

The unavoidable restlessness of middle world power

So the Bioindustrial Revolution went further than our human Industrial Revolution. It not only found a way to convert energy into work, it solved how to do it in the presence of unavoidable randomness.

We glimpsed in Chapter 1 how another protein engine, myosin in muscle cells, seems infected by randomness. With optical tweezer experiments on single myosin molecules, Toshio Yanagida and his team in Osaka showed that feeding a myosin some ATP did not result in straightforward engine motion. A myosin molecule might take one, two, five or no steps at all along its neighbouring actin fibre – it might even step backwards.

The indication from this and a host of other experiments is clear: life hasn't actually solved the restlessness problem at all. And yet the engines function. The solution that life's engines found to the randomness problem was not to fight Brownian motion, but to use it.

The answer is, as so often in the middle world, a question of statistics. A question, in fact, of *biasing* the statistics – of tilting the odds.

Manipulating the odds of randomness is nothing particularly special. Randomness doesn't have to mean entirely random: take bookmakers and casino owners. They live, undeniably, in a world of randomness. The lottery of the racetrack – the chance roll of the ball around the roulette wheel. And yet bookmakers and casino owners usually make a profit. They play a statistical game.

You can't take the fluctuation out of gambling – occasionally a zero has to come up on the roulette wheel, and occasionally a clapped-out old nag at 150-1 will get a surprise second wind and win the 3:30 at Newmarket. If not, who would bother gambling? But the bookies tilt their odds: they adjust the returns they're willing to offer to ensure that, on average, they win out over the randomness. They bias the game so that, despite the occasional big win, the flow of cash is, statistically speaking, out of punters' pockets and into theirs.

Life, according to the latest ideas of how engines like kinesin and myosin work, does the same: it biases the game. The randomness of Brownian motion is unavoidable. But the chemistry of the kinesin–tubulin or myosin–actin interactions tilts the odds just enough in the engine's favour that, on average, it can turn energy into directed motion.

In 1994 Dean Astumian and Martin Bier, at the University of Chicago, published a stripped-down theory of such a 'biased randomness' engine. They considered the motion of a simple particle, subject both to the random Brownian forces and to a separate force that periodically switches on and off. Their inspiration came from what was known about engines like kinesin and myosin: the on–off–on force is a simplification of the way the kinesin foot's energy level changes as ATP binds, reacts and unbinds again. As ATP binds, the strong attractive force between the foot and the tubulin switches off – the foot lifts. As ATP turns into ADP and unbinds, the foot–tubulin force switches back on, and the foot is stuck again to the track.

Astumian and Bier's foot–tubulin force, however, was not quite so simple. They imposed a *biased* attraction on the foot – essentially stronger in one direction along the track than in the other. When detached, subject only to random Brownian motion, the foot is indeed equally likely to wander backwards as forwards. But the biased foot–tubulin attraction encourages forward wanderings. A foot that wanders the wrong way tends to get dragged back to where it started, while a foot that wanders forward is allowed to continue unmolested.

Hence Astumian and Bier showed that positive directed motion is indeed possible even in the presence of Brownian motion – as long as the periodically switching foot–track force is biased to favour one direction over the other. Random odds must be tilted in favour of the engine.

How is this bias realized in practice, in protein engines like kinesin and myosin? It is the chemical structure of the tubulin track, or the actin fibres in muscle, that produces the equivalent of Astumian and Bier's directionally biased force. The interactions between the trucks and the tracks generate directionally biased forces, with the result that forward motion is encouraged relative to backward. On average, over many steps, it works out easier for the kinesin foot or the myosin protein to crawl along in one direction than the other.

It remains very much a statistical game. At any given step the engine's moving part, subject to the ravages of randomness, may still go forward or backward. But overall, the probabilities of the forward and backward directions are not quite the same, the chemical bias being enough to favour forwards.

As Toshio Yanagida's experiments showed, occasionally myosin does step backwards. Occasionally it even manages to dance a lot further than one step: such is the nature of randomness. Similarly, in Karel Svoboda's optical tweezer experiments on kinesin, the engine sometime steps backwards, but manages ultimately to make its way forwards. Only when averaging over a long enough time or number of 'engine cycles' does the overall one-way engine function become apparent – like a pattern that slowly emerges out of fluctuating static.

Astumian and Bier compared their simplified theory directly to Svoboda and Block's kinesin experiments. The predictions were in good agreement with the observations – remarkably so, given the simplicity of the theory versus the complexity of the chemical processes involved in the real protein engine.

Experiments on kinesin also show how increasing the 'load' on the engine critically affects its performance. If a big enough pull is exerted on the molecule via optical tweezers, a 'stall force' is reached where the load overwhelms the statistical bias in the foot–track interaction. The engine doesn't stop, because it is still converting energy via reactions with ATP and still subject to Brownian motion. Instead, its walk becomes entirely random. Efficient directed transport breaks down.

Similar optical tweezer force experiments have been done on other motors, such as protein machines that read and copy DNA and others that help two strands of DNA 'zip' back together into the familiar double helix. Again, these motors 'stall' if the load gets too great. Life's engines seem to have been tuned by evolution to fit function as optimally as possible to environment – not too dissimilarly to the expert tuning of a high-performance racing car[7].

---

7   Current experiments and theory are concerned with measuring just how efficient protein engines are and how efficient they *could* be. How close has evolution come to Sadi Carnot's thermodynamic limit on the perfect engine?

Collective action

Tuned or not, this 'two steps forward, one step back' sort of motion inherent in bioengines still sounds rather inefficient. Even with the requisite bias, it would be a painfully jittery way to drive to the shops. But there are two differences between a car engine in the macroworld and bioengines in the cell. Firstly, life's engines have *no choice* but to deal with randomness.

Secondly, the cell contains a huge population of molecular engines. Transport isn't a matter of one kinesin carrying one bit of cargo from A to B. A typical cell has an entire fleet of kinesin cargo trucks, thousands of bioengines, marching around the tubulin network. Some of them do go backwards now and then, some of them detach completely and wander off; but enough make it to their destination to keep the system going. Out of a thousand molecules, say, perhaps six or seven hundred will go in the right direction. So the 'statistical supply chain' continues and the cell, despite inherent jitters, survives.

Similarly, a muscle is a whole bundle of actin fibres and thousands of myosin molecules. Again, as long as a majority pull in the right direction, the muscle works: the actin fibres contract; I type the next word, you turn the next page.

Such collective action even opens up technological possibilities. Scientists are studying how armies of engines like kinesin may be used to build middle world structures. The statistics of how such a crowd of engines behaves may provide ways to carry out so-called 'self-assembly' of microscopic environments: set a crowd of protein engines going and their collective behaviour will result in automatic construction. The generation of the tubulin scaffold of the cell may itself be a case of such 'statistical teamwork' by protein engines[8].

---

8 This statistical or 'collective' behaviour, of everything from molecules to people, is currently the subject of huge research effort in fields ranging from theoretical physics to biology to sociology. Crowds, as is well known if not well understood, behave in ways that cannot easily be inferred from single individuals. See Philip Ball's recent popular book *Critical Mass* for many more examples.

Life – a catalogue of bioindustrial miracles

All proteins are engines of a kind. Another example is RNA-polymerase, an enzyme that reads a genetic sequence from DNA and makes a chemical copy, a molecule called RNA. (RNA is then transported out of the cell nucleus to the ribosome, where yet more enzymes – i.e. yet more bioengines – use it to manufacture new proteins.) To copy DNA, RNA-polymerase crawls along the DNA strand, assembling new chemicals in the required sequence. In binding to energy molecules such as ATP, the enzyme switches between different shapes. Its shape controls which chemical binds as the next ingredient in the new RNA chain.

All this shape shifting and chemical assemblage happens in the presence of continual bombardment from water molecules, the RNA-polymerase engine both coping with and using the fluctuating nature of the middle world.

Important engines also reside in the membrane of the cell. Some act as pumps to suck in salts and ions from outside, and some as waste disposal devices to get rid of unwanted cellular detritus or poisons. One multitasking engine-enzyme, called F_0F_1-ATP synthase, works in two ways: when ATP is plentiful it acts as a pump for nutrients across the membrane; but when there isn't enough ATP it works backwards, synthesizing more ATP. The engine F_0F_1-ATP synthase is thus key to the metabolism of the cell, balancing the manufacture and use of ATP.

In 1997, a team led by Masasuke Yoshida of the Tokyo Institute of Technology first directly visualized a working F_0F_1-ATP synthase engine. They showed that it is a kind of rotary motor. ATP-powered rotation of a piece of the protein inside a 'barrel' formed by the rest of the molecule enables the pumping of ions across the cell membrane.

Yoshida's team glued molecules of the motor onto a glass plate and attached fluorescent filaments of actin, big enough to be seen in the microscope, to the other end of the motor. When fed ATP, the tiny filaments started to spin round at up to four revolutions per second. Yoshida often saw hesitation in the rotation: rotors would pause, wait, start again, and sometimes even reverse direction. Another restless Brownian motor.

There are other rotary motors that function as biopropellers, enabling whole bacteria to swim around at speeds of up to a tenth of a millimetre per second[9]; motors that viruses use to pack themselves with DNA and to drill entrance holes into their unsuspecting victim cells; engines that build or deconstruct the cellular tubulin scaffolding; engines that unzip the two strands of DNA in preparation for copying or cell division; engines that repair mistakes and damage to the DNA sequence; engines that enable the transport of unfolded proteins through membranes.... The cell is, in short, a thriving factory containing a huge variety of industrious bioengines.

However minimal our understanding of this bioindustrial miracle is as yet, what we have discovered so far seems to be spectacular evidence of a momentous event in Earth's history. The Bioindustrial Revolution was not just about finding a way to convert energy into work, nor simply how to deal with randomness. It was about finding a solution to the energy-into-work problem at the same time as profiting from the randomness unavoidable on the scale of the middle world.

Just as the activity of proteins like myoglobin is, as we've seen, a fine balance between the rules of chemistry and the randomness, so the Bioindustrial Revolution was a productive compromise between the resources of chemistry and the possibilities opened up by middle world fluctuations.

The rules of chemistry alone are too rigid; randomness alone too undirected. Without one or the other – rules *and* restlessness – engines of life couldn't even have got into first gear, let alone

---

9   For a bacterium of, say, a thousandth of a millimetre in diameter, this speed is something like a two metre long car moving at 200 metres per second – 720 km per hour.

accelerated away to turn a lifeless rock into the teeming, living globe we call Earth.

The human Industrial Revolution left behind plentiful reminders, from vast textile mills to steelworks to railways to coal and tin mines. Many have become crumbling ruins, echoing and empty, weed-grown and abandoned.

The remains of the Bioindustrial Revolution are neither crumbling monuments nor weed-infested ruins. The ubiquitous, and still very much functioning, relic of the Bioindustrial Revolution is the cell. Cells everywhere, from bacteria to plants to protozoa to animals, billions upon billions of examples of organized, busy, efficient, self-contained industry.

And all of it depending fundamentally on how matter behaves in the middle world – all of it powered by middle world engines.

Chapter 12
MASTERS OF MIDDLE WORLD

It's time to talk technology.

Annual global research and development spending on middle world technology has already reached $9 billion, according to a December 2005 news article in the journal *Science*. By 2015, predicts the US National Science Foundation, middle world technology will have a $1 trillion 'footprint' and employ two million people.

Can all this money really be going on middle world technology? Who has ever heard of middle world technology?

No one has: it is otherwise known as 'nanotechnology'.

A nanometre is one billionth of a metre. A common or garden ant (UK style) is about five million nanometres long – five millimetres. The smallest scale the human eye can resolve is around a tenth of a millimetre. An object this size would be one hundred thousand nanometres long.

Nanotechnology is the technology of making things that are from a few nanometres to a few hundred nanometres in size. An atom is around a tenth of a nanometre[1]. So nanotech objects are made up of a few hundred to a few thousand atoms.

In other words, nanotech objects are middle world objects.

A recent explosion in middle world technology has come about thanks to new ways of observing and manipulating tiny objects – new toys like the scanning electron microscope, the atomic force microscope, and, as we've seen, microscopic manipulators based

---

1 The size of an atom is not a straightforward quantity. It is best thought of as the scale at which the atom 'feels' neighbouring atoms.

on light-traps. At last, it seems, scientists and industrialists may be on the verge of mastering the middle world.

It would be a mistake, though, to imagine that this technology is all new. In the early 20th century, plastics engineers first used the potential of the new chemistry, in other words polymer chemistry. Middle world technology goes back a lot further even than that. The ancient Egyptians were amongst the first middle world pioneers.

Middle world writing

Around 5000 BC the Egyptians invented papyrus, the original form of paper, as a more convenient alternative to carving hieroglyphs on wax tablets. To use paper, they also needed ink.

First tries with suspensions of charcoal dust in water ran into a shelf-life problem. Too quickly the charcoal particles clumped together and sank to the bottom of the inkpot, resulting in unusable gunk. How to make ink that could last longer?

The Egyptians found that adding gum arabic, a tree resin, to the charcoal–water mixture stopped the charcoal particles from coagulating together, vastly extending the use-by date of the ink. They had little idea why such a trick worked. The explanation had to wait for a modern understanding of polymers.

Gum arabic resin consists of polymeric molecules – carbon-based chains. Added to a charcoal–water ink, the gum arabic polymers attach chemically to the charcoal grains, giving the grains something like a head of polymer hair.

However, the links of the gum polymer chains are also attracted to the surrounding water molecules. So the chains want to achieve maximum contact with the water molecules. The polymers want to be 'swollen' like the unfolded proteins we've already seen. So each charcoal grain is clothed in a swollen, incessantly wriggling coat of gum arabic polymers. (In terms of a hairstyle, it's more of an afro than a crewcut.)

Should two of these polymer-coated charcoal grains come close together, their writhing polymer coats start to entangle. The polymers' overwhelming desire for freedom to wriggle and remain in maximal contact with water molecules means that such an approaching entanglement results in a strong repulsive force between the two nearby particles. The charcoal grains are pushed apart by their polymer coats.

If the grains can't get close together then they can't stick – hence the ink doesn't coagulate into lumps. Though the ancient Egyptians didn't know it, their improved ink technology was making use of the key feature of the middle world, that objects like polymers can't stand still.

According to the Roman writer Pliny, users of this engineered Egyptian ink ran into another problem: forgery. Once written onto papyrus the gum-treated ink was alarmingly easy to rub off. A fix was found, Pliny tells us, by adding vinegar to the ink.

Chemically speaking, the polymer-coated charcoal grains were not only reluctant to stick to each other, but also to the papyrus. The acidity of the added vinegar encourages the charcoal particles to stick better to the papyrus, without making them coagulate in the ink solution. More chemical engineering in the middle world.

In an echo of ancient technology, similar acidification was used by Jean Perrin, in 1908, to encourage his *gamboge* particles to stick to his microscope slides, so that he could get better size measurements. Perhaps Perrin had read Pliny.

Making plastic less primitive

The plastics we've already met are one example of 'middletech'. The next step beyond these simple polymers is a class of molecule called the 'copolymer'. Typical copolymers are still hydrocarbon chains, but they are one step up in complexity from your basic run-of-the-mill plastic. Copolymers are made up of alternating units or sub-chains that have distinct chemical characteristics.

The different 'sub-species' in the copolymer are distinguished by the differing character of their chemical interactions – interactions either with other parts of the chain, or with the surrounding solvent molecules (e.g. water molecules if the copolymer is dissolved in water). It's Paul Flory's balance of interactions again – but now, due to the difference in chemical characteristics from sub-chain to sub-chain, that balance is engineered to achieve some extra control over polymer structure.

For example, consider a copolymer made of two types of sub-chain, one water-repellent, one water-attractive. A copolymer that is 80% water-attractive sub-chains will, overall, like being in water and will dissolve relatively easily. Change the percentage so that the water-repellent type is in the majority, and overall the copolymer will tend to dislike water. It will be difficult to make it dissolve.

So, at the simplest level, how the polymer behaves in water (or other solvents) can be engineered simply by changing the relative amounts of different sub-species.

This ratio of different species also partly determines the shape of the copolymer in solution. Say we dissolve the 80% water-attractive copolymer in water. To minimize its contact with water molecules, the water-repellent part crumples up in the middle, shielded from the water molecules it hates by the surrounding water-attractive part of the chain.

Whether such a shape is possible will also depend on the order of the sub-species along the copolymer chain. If all the water-repellent species are at one end of the chain, they can easily crumple up and be surrounded by the water-attractive end wrapping around them. If they are spread throughout the chain it might not be so easy.

So a chemist manufacturing a copolymer from two sub-species gains a lot of control over the shape and chemical behaviour of the product, both by balancing the relative amounts of different species and by chemically directing the order in which they get chained together. For example, you could have a chain made from two species, A and B, arranged in a simply alternating sequence, ABABABA.... Or what about something a bit more complex: AABAABAABAABAABAA.... Control of the sub-chain sequence opens up a world of chemical shape and structure.

And of course you don't have to stop at two species: 'tri-polymers' are made from three polymers, which could be arranged in order, ABCABCABC... or even nearly randomly, such as AABBAACCAACAABBAABAAC.... As the number of sub-species increases, the range of possible combinations, sequences and shapes mushrooms. Since the early days of synthetic polymer chemistry many thousands of markedly different polymers have been made[2].

Proteins are essentially an advanced sort of copolymer. The sub-species are the amino acids, of which there are 20 distinct types found in natural living systems. Twenty is a big number when it comes to figuring out the total number of possible sequences thousands of amino acids long. In proteins, it is the sequence of amino acids that determines the interactions between pieces of the chain (as well as between chain and water and chain and other molecules), and these interactions deter-mine the shape of the protein, which as we have seen is all-impor-tant for its function. Copolymers really do have vast potential for complex chemical function, even to the point of life itself.

The first attempts to master the middle world of proteins for technological purposes have been made not so much by chemis-try but by fiddling with biology. A biochemist comes up with an amino acid sequence that might make a particular interesting structure, then inserts sections of DNA into a bacterium to repro-gram it so that it builds the required new protein. For instance, in early 2006 a multinational team led by Hak-Sung Kim in Korea altered an enzyme-protein to have a completely different func-tion from the original. Life's machines re-engineered.

Tops and tails and sponges and worms

A special class of copolymer is the surfactant, short for 'surface active agent'. Surfactants are chains of carbons with a top and a

---

2   Only about a hundred or so of these manufactured polymers have ever been exploited in commercial products.

tail. The top, usually a part with an electric charge, is attractive to water, while the tail, a part mostly oily hydrocarbons, is water-repellent.

Surfactant molecules, beyond being middle-worlders themselves, form a host of collective structures that are also middle-world-sized, from spheres to flat layers to tortuous sponges to closed membranes. These are all dynamic, flexible, wriggly objects with great potential for technologies from the prosaic – such as the washing-up after dinner – to the sublime – hi-tech targeted drug delivery.

Prosaic first. When both water and oil are present, such as on your greasy dinner plates in the washing-up bowl, surfactants adopt a collective behaviour. To best satisfy the water-repulsion of the tails and the water-attraction of the tops, the surfactants arrange themselves with their tops in the water and their tails in the oil, collecting at the interface between water and oil. Hence 'surface active agent'.

This flocking to the water–oil interface can be best achieved if the oily coating on the dinner plate is lifted off to form bubbles of oil immersed in the water. Each oil bubble is coated by surfactants with their oil-bound tails inward and their water-bound tops outward. Cleaning processes are basically a technology to encourage water and oil to mix[3].

Sublime next. One of the potentially most exciting developing uses of surfactants is in the delivery of drugs into the body. Surfactant membranes can be used as wrapping to shield 'alien' drugs from the outer defences of the cell or organism. Chemicals can even be incorporated into the membrane to 'target' the contents to the right location in the body. Once the drug has reached its target, the surfactants dissolve away to let the drug go to work.

But we can't leave this whistle-stop tour of middle world surface technology without mentioning shampoo.

---

3   And salad dressing too: surfactants in mustard help the oil and vinegar (acidic water) mix, ready to be poured on your lettuce.

The next time you're in the shower getting a good lather going, consider this: your shampoo is full of worms.

Perhaps I should say 'worms', or to give them their proper title, 'wormlike micelles'. These are surfactant assemblies where the molecules collect together to make long flexible cylinders, with their tails pointing into the inside of the cylinder and tops on the outside. These cylinders writhe and entangle similarly to polymers, but unlike polymers, whose chain links are strong chemical bonds, the surfactants in a wormlike micelle aren't really tied together. As well as wriggling and twisting worms also split into pieces and join other worms.

The worm shape is ideal for shampoo and hair conditioner, since these worms can wrap around strands of hair, clinging on for long enough to maximize the delivery of detergents and conditioning oils to the hair.

The next time you do the washing-up, or make a salad dressing, or step into the shower – look on it as a journey into the middle world.

Nanomachines: spinning and swinging in the middle world

Shampoo and salad dressing are all very well, but probably still quite distant from most people's idea of a true nanomachine – some powered middle world robot busily executing its program, constructing, deconstructing, transporting or indeed defending and attacking (for instance fighting a virus or a cancer cell). Recent achievements in such hardball nanotech indicate that the dream of artificial, controllable middle world machines might not be so far away.

For instance, in 2005 a team led by Ben Feringa at the University of Groningen in the Netherlands made a rotary molecular motor driven solely by chemical reactions – reminiscent of, if still far less sophisticated than, life's rotary engines such as the ATP synthase protein. The Groningen team's motor consists of a molecular structure in two flexible parts joined by a single carbon–carbon bond. Chemical reactions rotate one half relative to the other. The shape of the molecule means that once rotated the twisted half can't just rotate back again. Hence chemistry generates unidirectional rotational motion.

But don't sell the car yet. The reactions that drive the rotation come in four discrete steps, each involving different chemical additives. The chemist has to push the motor around by adding the right stuff at the right time. Not yet a very practical nanomachine.

Rotary middle world motors have been devised that work by light, also by Feringa's group and by others such as David Leigh's team at the University of Edinburgh. Molecular rotation is triggered by laser light – a more automation-friendly process since semiconductor laser switches could be incorporated into electronic control circuits.

Spinning isn't everything: in 2005 another group of researchers, this time led by Jannik Meyer of the Max Planck Institute in Stuttgart, Germany, made not a nanorotor but a nanopendulum. This came in the form of a carbon nanotube – a cylinder made out of pure carbon, about two nanometres in diameter – onto which the team hung a tiny mass of metal (less than one ten-thousand-billionth of a kilo). The experiment demonstrated how such a nanopendulum is at the mercy of the middle world: bombarded by the surrounding gas atoms it swings back and forth in random Brownian motion.

These nanomotors are based on bottom-up chemical invention. The cyborg route to middle world engines may be a more reliable route to success. After all, evolution has spent the last four billion years or so developing the base technology. In 2000 a team led by Carlo Montemagno at Cornell University swallowed their pride and employed life's ready-made rotary motor, the ATP synthase protein, in an artificial nanodevice.

The team stuck the protein engine to a glass plate, and fixed a nickel 'nanopropeller' about a thousandth of a millimetre long to the rotor of the engine. Feeding ATP to the cyborg resulted in the propeller rotating. Part bioengine, part metal – the Industrial and Bioindustrial Revolutions beginning to come together.

The hard and soft options

What of the middle world and Brownian motion, amidst all this talk of light-driven nanomotors and cyborg engines?

Much of the scientific and commercial interest in nano- or middle-world-technology can be traced back to Eric Drexler. In 1986 Drexler wrote a book called *Engines of Creation* in which he prophesied the new technological revolution: a world driven by swarms of microscopic machines doing nanotasks, everything from atom-by-atom materials construction to cell-by-cell medical repairs.

Much of the inspiration for this dream can probably be traced to the electronic revolution, whose main drive was the seemingly limitless advance of miniaturization. In the 1950s when the transistor was first produced it came in the form of a metal can a few millimetres on a side, with wires coming out of it like little legs. Today's microprocessors pack millions of such devices into the same size space.

Nanotech Drexler-style was a logical extension of the miracle of miniaturization: why should miniaturization be limited to making electronic circuits, why not do the same thing with matter in general and construct all kinds of devices at the microscopic scale? If digital electronics could revolutionize the world simply through its effect on information processing, imagine the even greater revolution that would come from using the same miniaturization concept, but in a far vaster range of operations, from device construction to medicine.

However, another name for the nanoworld is, of course, the middle world. As many scientists have pointed out since Drexler, simply taking the things we do up here in the macroworld and shrinking them into the middle world isn't quite going to work.

In his book *Soft Machines*, Richard Jones, a nanotechnologist at the University of Sheffield, gives an apt illustration of how the behaviour of middle world matter will force nano-engineers to change the way they do things. Imagine one simple everyday macroworld machine: the bicycle. Its mechanisms are relatively simple: rotate one part with your feet and that rotation is fed through gears to the back wheel, which, by friction with the ground, drives the whole contraption along in a straight line.

So what about shrinking the bicycle into the middle world and calling it the nanobike? The nanobike could, because of its simplicity, revolutionize the way cargo is transported down there. Where kinesin walks, we shall bicycle.

The problem is Brownian motion. As Jones points out, the nanobike would wobble and bend and wriggle and shake – all under the bombardment of surrounding atoms – not much different from a polymer molecule in plastic, or indeed the moving parts of kinesin. Try riding a bike that's about as solid as custard.

Well, argues the would-be nanotechnologist, make it tough! Make it out of molecules that have such strong interactions that the damn thing doesn't bend!

There are ways to make a rigid nanobike using strong chemistry – just as a protein, for instance, in the right conditions can more or less keep to a particular conformation. However, even if your hard bike worked, would it really work very well in the restless middle world? Besides its inherent difficulties, the hard option – as most nanoscientists these days have realized – misses a big opportunity.

Matter in the middle world does things differently. You could insist on modelling your machines after the macroworld and finding chemical ways to achieve that. But why not use the fantastically rich range of things that matter does in the middle world to come up with whole new ways of solving engineering problems? Why not profit from unavoidable restlessness? We know it can be done: life has already done it.

This is more or less preaching after the fact. All the examples of middle world technology I have talked about in this chapter take the soft option in some degree or other. Polymer-stabilized Egyptian inks, proto-protein copolymers, wormlike shampoos and ATP-powered cyborg propellers all make intelligent use of how matter behaves in the middle world.

In fact successful middletech seems likely to take the hard-and-soft option. Exploit the balance between the different dominant characteristics of matter on different scales, from atomic and molecular chemistry, to Brownian middle world fluctuations, to macroscopic structure. No one scale needs to be predominant. The middle world is the bridge between scales of reality, from carbon atoms to shampoo to salad dressing to life itself.

Sharing our middle world

But technology has never been without its dangers, and middle world technology is no exception.

Is nanotechnology a threat? A few years ago, none other than Prince Charles, heir to the British throne, issued a warning: were we going to sit back and wait for the 'grey goo' of microscopic machines to eat us up atom by atom? Others with perhaps better scientific credentials are also worried. In April 2000 techie-deity Bill Joy, founder of computer company Sun Microsystems, called on scientists to stop doing nanotechnology: it was just too risky. At the end of 2005, top technology companies, including Dupont, along with environmental pressure groups, the UK Royal Society and the Japanese Science Council, clubbed together to call for more research into the health and safety implications of nanotechnology. They pointed out that from $1 billion of annual US Government research funding in nanotechnology, a paltry $39 million was dedicated to investigating health and safety.

Why should nanotechnology be any worse than any other sort of chemistry, or manufacturing, or many other industries?

There is at least one reason why nanotechnology or middle world technology might present greater dangers to us than other industries. The middle world is the predominant scale of the mechanisms of life, of proteins, enzymes, cells, cell membranes, DNA, RNA and molecular motors.

Nanotechnology is playing in the same playground as cellular life, utilizing the same middle world processes. Playing around in the middle world means, inevitably, playing around in the world of life.

While this has obvious potential for catastrophic mistakes, it is also one of the greatest potential advantages of middle world technology. By learning about the mechanisms of life, such as the way protein engines efficiently convert energy to motion, we may be able to develop technology that helps solve some of life's most fundamental problems. Nanomedical dreams such as cell-by-cell cures and treatments for cancer; new ways to protect cells from viruses and bacteria; and technological replacements for failing organs and malfunctioning cells. Even, perhaps, the ultimate challenge: combating ageing and cell death itself.

But we certainly don't yet understand anything like the full range of rich possibilities of what matter can do in the middle world. This makes it seriously difficult to predict possible unwanted interactions between our own invented technology and the living, natural stuff that is already there. In trying to help life in the middle world, we may accidentally hinder it instead.

So we should certainly be concerned about nanotechnology. The engineering potential of the middle world has been exploited by humans since at least 5000 BC. As our understanding grows, we are beginning to learn just how fantastic that potential is. And yet... we are part of the middle world in a way that we are not part of the world of any other technology.

We are latecomers as middle world engineers. Life got there first. We *are* middle world machines: we have to be careful that we know what we are doing, before we let loose too many new ones.

They may not be entirely happy to share the middle world with their creators.

Chapter 13

THE RESTLESSNESS OF
MATTER AND LIFE

We have travelled a long way since that fateful summer's day of 1827, when Robert Brown first peered through the lens of his microscope and glimpsed the middle world. This great yet nowadays neglected Scots botanist was the first to make real landfall in the new world of restless matter.

Since then science has gradually – sometimes almost frustratingly slowly – learnt more about this peculiar place and its odd inhabitants. And it was, we can see now, inevitable that scientists should have to come to terms with what can happen in the middle world. Because without such understanding, theories of everything from heat to atoms to stock markets to proteins to muscles to nanotechnology are simply impossible.

For all that, the middle world even now retains a relatively low profile in the popular image of science. Isn't science all about quarks and galaxies? I hope I've convinced you that there's something very important happening here, in the middle world between the size of quarks and the size of galaxies. Without understanding the middle world, we've no chance of comprehending a huge range of fundamental phenomena going on around us and inside us every minute of the day.

There's a nice irony at the heart of the middle world story. Robert Brown began with the intention of studying life, as he embarked on those experiments in 1827: he was, after all, a botanist. He set out to examine the microscopic function of pollen – the scientific basis of plant sex. Stumbling on the strange random dance of the middle world, he used careful experiments to demonstrate that, whatever it was, this bizarre behaviour was not due to some weird 'life force'. So – or so he thought – the problem was shifted squarely onto the table of the physical scientists. This restlessness he had found was a question of matter, not of life.

Around three-quarters of a century later physicists finally explained the random dance of the pollen, and by measuring it they proved that atoms are real. Matter is made out of broken pieces.

Another three-quarters of a century later, give or take, advances in experimental optics, lasers and cell biology began to demonstrate how the track of middle world science was turning back on itself. DNA and proteins and molecular motors and bacteria and viruses – all these are middle world inhabitants too.

What Robert Brown saw and wrote off as none of a botanist's business came around full circle. The random dance of the middle world was very much the business of the biological sciences after all. It was a question of life as well as matter.

Life in the middle

The story of the middle world could be divided roughly into two parts: matter and life. But such a clear distinction would be largely artificial. In the middle world, the gap between matter and life is more a blurred region than a sharp border – polymers, surfactants and membranes shade into proteins, molecular motors and cells. If, 50 years from now, we look back, I think we'll see the tale in a more singular light.

We'll see that all along the science of the middle world has really been about understanding the common ground between matter and life. It is – has been, will be – the story of how scientists grasped that intimate relationship; how they grasped the full and surprising range of what matter can do, and how they grasped that *life* is one of the most special things that matter can do.

Right here today we're a long way from the end of the story: many of life's processes that root our existence in the middle world remain so complex that they could still *almost* be miracles. But in only the last 20 years we have come a long way: there are many signs that the gap between matter and life, a gap securely located in the middle world, is beginning to close.

Perhaps such a 'material' concept of life – life as a function of matter – seems a bit disappointing – frightening, even. Is life just another 'mere' function of matter, like gravity and static electricity and table salt? Doesn't that take some of the specialness of life away?

I would say quite the reverse. What the middle world is showing us is just how special life really is – indeed, just how special matter is. Physics – the science of matter – has long had an image dominated by reductionism: reducing the whole of chemistry to a few quantum calculations; reducing the whole of electricity, light, magnetism, radioactivity and nuclear glue to the Standard Model of the fundamental forces. If physics could go ahead and 'reduce' life to some *simple* function of matter, I too might side with the romantics and shake my head in disappointment.

However, the functions of matter in the middle world are *not* simple. They never will be, however clever we get. It's a complicated place. What scientists are revealing most of all today is just what an astounding range of clever and complicated things matter is capable of. Including life.

Middle world science isn't about *reducing* life to a function of matter: it's about raising the functions of matter into life. It isn't about reducing biology to physics: it's about joining biology and physics.

In the science of the middle world we can start to see the old divisions between science disciplines weaken. Read a research paper on molecular motors or enzymes or nanomachines in the pages of journals like *Nature* or *Science*, and you will hardly be able to see the joins between the biology and the physics and the chemistry. This is simply the science of what matter can do.

What else will we see when we look back at the story of the middle world in 50 years time? Undoubtedly lots of the details of life-processes will have been worked out; and just as undoubtedly, many of those details, and many new mysteries, will still be puzzling us. We may by then have learnt enough from middle world life-machines to make our own realistic nanomachines – and we may even have managed to do it safely and in a way that provides some benefit to those who really need it – instead of just more over-elaborate gadgets and unforeseen dangers.

Hopefully, the last vestiges of that blind leap, the original Newtonians' assumption that only the planets and the atoms mattered, that there was nothing of interest in between, will have been eliminated. Hopefully, via the bridge of the middle world, we will be able to see the functions of matter – from subatomic quantum weirdness to atomic chemistry to middle world dancing to the macroworld's complexity – as a continuum: not a hierarchy where one scale is more important than another, but a jigsaw where everything fits rather beautifully together.

But if I had to choose one question that I would really like middle world scientists to throw more light on over the next decades, it's this: the nature of the Bioindustrial Revolution. The origin of life.

Brownian soup

The bits and pieces that make up the mechanisms of life, as we have seen in this book – the molecular engines, the tubulin transport networks, the proteins, the membranes, the enzymes, the DNA – are all inhabitants of the middle world.

The Bioindustrial Revolution was perhaps the most significant event to date in the history of the middle world. This was when, somehow, material objects appeared that achieved that great ambition of our own industrial pioneers: to turn energy into useful work. Before that revolution, it was a case of 'energy, energy everywhere' – but not a drop that could be used to do anything with. Once those engines of life had begun to work, suddenly the great harvest of resources could begin.

There exist a rather large number of more or less plausible theories about how and where life started. This is not the place to go into them (see for example Paul Davies' book *The Origin of Life*). Rather, all I want to do here is ask a question:

Could this Bioindustrial Revolution have occurred any-where but in the middle world?

Perhaps we still don't know enough about matter to be sure: perhaps there are other things, on other scales, in other environ-ments, that matter is capable of – things that we might call life, or might rival life in their complexity.

We certainly can be sure, though, that in our case, here on Earth, the middle world *was* the place that the processes that eventually led to us settled on. The best workshop for construct-ing life's first industrial machines.

It seems most likely, in other words, that terrestrial life – in the form of the first bioengines able to convert chemical energy into useful work – began in the middle world – whatever form that 'beginning' took. Which means that when it comes to the origin of life, even if we don't yet know exactly *what* to look for – perhaps we do know *where* to go and look for it.

And we know something else. Brownian motion was there. Wherever and whatever it was, the primeval soup was Brownian-flavoured.

The inevitable dance? – A last speculation

Whatever events led to the first identifiably living processes, the restlessness of middle world matter must have been intimately involved. Perhaps these processes included primitive Brownian engines, molecules something like kinesin, the first glimmerings of a cargo-transport system. Perhaps rippling membranes of sur-factants, the first sign of cellular partitioning. Perhaps ancestral copolymers, the first step toward that balance of chemistry and randomness that enables structured molecular machines. Per-haps even early versions of DNA-copying enzymes, the first incarnations of machines vital for memory and evolution. Again, theories abound, and only more understanding of the middle world and the details of the Bioindustrial Revolution will help us decide between them.

The subsequent evolution of life – producing a seemingly endless proliferation of ever more complex ways to take advantage of the nature of matter – must in its turn have been profoundly affected by restlessness. Evolution is essentially an exploration of the possibilities of matter: matter's way of finding ever cleverer ways of using the tools at its disposable. In many ways this evolution, this exploration, is enabled by the way matter does things in the middle world. Mistakes, for instance: they don't happen without randomness. And without mistakes in the copying of DNA, evolution doesn't happen. Perhaps our molecular memories *need* to be stored in the middle world – where evolution through randomness and mistakes can be finely balanced against fidelity of memory through the rules of chemistry.

But there is a further possibility that goes beyond even this. Now I really *am* speculating – let's underline that. But it's a powerful enough *what if?* to make it worth mentioning all the same.

What if life is an *inevitable* consequence of the middle world?

Imagine, for example, a primitive superplastic, a chain of amino acids. Once the chain has grown to middle world size it can't avoid wriggling around. So this piece of matter is drawn, inevitably, into the blurred region between the rules of chemistry and the randomness of the middle world. Drawn within reach, perhaps, of the fine balance that would result, after millions of years of further evolution, in the fantastic precision and facility of a real protein or enzyme. Primitive superplastic, a product simply of chemistry and middle world fluctuation – but also a first glimpse of life?

In other words, what if, once you have wandered into the middle world of matter, in all its strangeness, its balance of chemistry and writhing, wriggling random restlessness – what if you cannot *avoid* lifelike functions appearing?

Well, a primitive protein is all very well. Can we extend this idea to the whole of life? Does all life's complexity arise naturally, inevitably, from superplastic? We are nowhere near being able to give a definitive answer to that, and it would be supreme folly to imagine that life doesn't have a huge supply of complete surprises still to deliver.

It would certainly make a very nice conclusion to this book, to the story-so-far of the middle world: a story that began with Robert Brown the botanist and his dancing pollen, his abortive attempt to study plant sex – and perhaps one day will end with a truly startling idea: that *Brownian motion causes life*.

Enough! End of speculation – for now at least. We shall just have to wait and see – we shall just have to go on exploring the fantastic complexity of the middle world, and, just like Robert Brown that June day in 1827, keep our eyes peeled for more surprises.

The middle world story

We are still nowhere near really knowing all the possible consequences of the middle world. We've seen a lot of them, from entropy and steam engines and ununstirrable rice pudding, to microscopic atmospheres of rubber balls, to plastic, to life's own power industries in the form of molecular motors. We've seen how some of the greatest scientists in modern history have been intimately involved, even when they didn't know it, in the story of the middle world – from Isaac Newton to Robert Brown, from Sadi Carnot to Rudolf Clausius, from James Clerk Maxwell to Ludwig Boltzmann, from Albert Einstein to Paul Langevin, from Leon Gouy to Jean Perrin. Not to mention the 'moderns', like Staudinger, Flory, Perutz, Frauenfelder, Chu, Ashkin… and a thousand other names yet to gather the crust of history and the apocryphal barnacles of anecdote. We have seen that atoms are real, we have discovered the true nature of heat and why it only flows in one direction; we have explored Gardens of Delight and Orchards of Determinism; we have seen how we owe our lives, on a cellular level at least, to the middle world balance between chemistry and randomness. We've seen how today's technologists are trying to master the middle world. We've seen how life seems actually to have done that already, way ahead of us, almost 4 billion years ago. That's about where we are today: still rather ignorant latecomers to this strange universe in-between.

We have even wondered if the middle world might hold the secret of how we came to be.

So we've come a long way and explored a lot of the middle world. But the story is far from over. I suspect that vast realms of the world in the middle remain to be discovered. Mysteries of the middle: still waiting to surprise, puzzle and enlighten us.

FURTHER READING

This is not an academic textbook so I have avoided detailed references in the text. Here I give a short list of books of general interest to anyone who wants to find out more about some aspects of the middle world, or about some of the scientists and science described here. This is hardly an exhaustive list, of course, and is in no particular order.

Richard Jones (2004) *Soft Machines*. Oxford University Press, Oxford.
A good introduction to the modern science of the middle world, especially concentrating on nanotechnology and, to some extent, biophysics. Includes more detailed science background on many of the topics in *Middle World*, but still written more for the general reader rather than the specialist.

D. J. Mabberley (1985) *Jupiter Botanicus: Robert Brown of the British Museum*. Braunschweig/British Museum, London.
About the only detailed biography of Robert Brown – and I do mean *detailed*. Heavy going and rather obscure, but a mine of Brown information, so to speak, especially on matters botanical.

Graeme K. Hunter (2000) *Vital Forces: The Discovery of the Molecular Basis of Life*. Academic Press, New York.
Very good on the people at the centre of the development of our 'molecular' understanding of biology – proteins, DNA, and so on.

David Lindley (2001) *Boltzmann's Atom*. Free Press, New York.
One of very few accessible books on Ludwig Boltzmann, for the general reader, with plenty of interesting biographical and historical information as well as a fair helping of Boltzmann's science.

Basil Mahon (2003) *The Man Who Changed Everything.* John Wiley, Chichester.

A well-written recent biography of James Clerk Maxwell, that strangely neglected 'national treasure'.

Tom Stoppard (1993) *Arcadia.* Faber & Faber, London.

Well worth a read – or a view, if you are lucky enough to catch a performance – a very funny, very clever play, with the mysteries of thermodynamics and time at its heart.

Mary-Jo Nye (1972) *Molecular Reality.* Macdonald, London.

The detailed story – written by a *real* science historian – of Jean-Baptiste Perrin's triumphant experiments in the middle world that proved the reality of atoms. Very comprehensive historical, philosophical and political background, especially in 19th century France, leading up to Perrin's work and 'proof of the atom'.

Philip Ball (2005) *Critical Mass.* Arrow, New York.

A compendious exposition of the role of statistics and 'collective behaviour' across a host of scientific disciplines.

Paul Davies (2003) *The Origin of Life.* Penguin, London. (Previously published as *The Fifth Miracle.*)

Much more detail on the complex chemistry of cellular processes, and on the plethora of theories of how it all got started.

Bill Bryson (2004) *A Short History of Nearly Everything.* Black Swan, London.

A sightseeing trip through a vast swathe of science and history; what can only be described as a highly entertaining scientific romp. Look out for numerous ghostly appearances by our old neglected friend Robert Brown.

Marcel Proust (2002) *A la Recherche du Temps Perdu.* Gallimard, Paris. (English translation C. K. Scott Moncrieff, e.g. Vintage Books 1982.)

Go on, have a read! Time will never seem the same again!

ACKNOWLEDGEMENTS

I must thank, of course, many scientists and non-scientists who have helped in one way or another in my own struggle to understand something about the middle world. I particularly acknowledge information and opinions from Charlie Laughton, Hans Frauenfelder, David Klenerman, Adrian Mulholland, Alan Cooper, Tom McLeish, Herman Berendsen and Peter Wolynes. As to history, I must thank in particular Dick Dougal, recently retired from the University of Edinburgh: a veritable fount of knowledge about Robert Brown and James Clerk Maxwell. And I certainly can't go without thanking all those at Edinburgh who contributed to my own scientific introduction to the middle world. It was, perhaps appropriately, in Edinburgh that I first stumbled on the middle world: if I have managed to stay there long enough to explore at least one tiny corner of it, it is only through a combination of support, encouragement and, yes, funding, generously provided by a whole variety of people and organizations.

Having thanked all these people, rest assured that any misunderstandings, errors and speculations are entirely my own.

Finally, for her endless encouragement and patience with me while I was lost incommunicado in the middle world, my unlimited thanks go to Eleanor.

INDEX

VENOMOUS EARTH
HOW ARSENIC CAUSED THE WORLD'S WORST MASS
POISONING
by Andrew Meharg
MACMILLAN; ISBN: 1–4039–4499–7 £16.99/US$29.95; HARDCOVER

"Meharg is good on the technological and political challenges of testing water. He is terrific on the wider history of arsenic, in alchemy, industry and interior decorating." *Guardian*

"Meharg tells the lively and cautionary story of arsenic's misuse over the centuries." *Newsweek International*

order now from www.macmillanscience.com

MAX BORN

THE
BORN–EINSTEIN
LETTERS
1916–1955

Friendship, Politics and Physics
in Uncertain Times

Introduction by **Werner Heisenberg** Foreword by **Bertrand Russell**
New Preface by **Diana Buchwald and Kip Thorne**

THE BORN–EINSTEIN LETTERS
FRIENDSHIP, POLITICS AND PHYSICS IN UNCERTAIN
TIMES
by Max Born and Albert Einstein
Introduction by Werner Heisenberg
Foreword by Bertrand Russell
New Preface by Diana Buchwald and Kip Thorne
MACMILLAN; ISBN: 1–4039–4496–2; £19.99/US$26.95; HARDCOVER

"An immensely readable personal account of Einstein's struggles with other physicists." *Washington Post*

"With a well-informed introductory essay by Buchwald and Thorne, the correspondence is a delight, enabling us to trace the development of the intriguing friendship between the two physicists and to read their views on the great themes of physics and politics of their time." *The Times Higher Education Supplement*

order now from www.macmillanscience.com

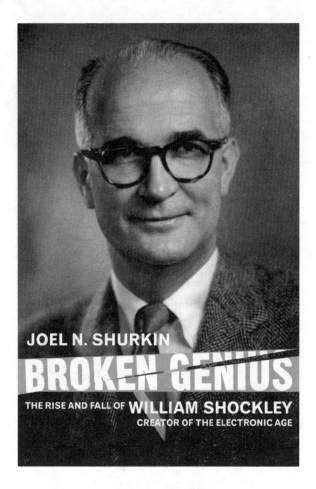